Geographical Data

Geographical Data Characteristics and Sources

Nigel Walford
Kingston University

JOHN WILEY & SONS, LTD

Baffins Lane, Chichester,
West Sussex PO19 1UD, England

National 01243 779777
International (+44) 1243 779777

e-mail (for orders and customer service enquiries): cs-books@wiley.co.uk
Visit our Home Page on: http://www.wiley.co.uk

Other Wiley Editorial Offices

John Wiley & Sons, Inc., 605 Third Avenue,
New York, NY 10158-0012, USA

Wiley-VCH Verlag GmbH, Pappelallee 3,
D-69469 Weinheim, Germany

John Wiley & Sons (Australia) Ltd, 33 Park Road, Milton,
Queensland 4064, Australia

John Wiley & Sons (Asia) Pte Ltd, 2 Clementi Loop #02-01,
Jin Xing Distripark, Singapore 129809

John Wiley & Sons (Canada) Ltd, 22 Worcester Road,
Rexdale, Ontario M9W 1L1, Canada

Library of Congress Cataloging-in-Publication Data

Walford, Nigel.
Geographical data: characteristics and sources/by Nigel Walford.
p. cm.
Includes bibliographical references (p.).
ISBN 0-471-97085-9 (alk. paper)
1. Geographic information systems. I. Title.
G70.212.W35 2001
910'.285'57—dc21 2001046626

British Library Cataloguing in Publication Data

A catalogue record for this book is available from the British Library

ISBN 0 471 97085 9

Typeset in 10/12 Baskerville by Vision Typesetting, Manchester.
Printed and bound in Great Britain by TJ International, Padstow, Cornwall.

This book is printed on acid-free paper responsibly manufactured from sustainable forestry, in which at least two trees are planted for each one used for paper production.

For Ann

Contents

Preface

Information and communications technologies have increasingly become the basis for analysing, handling and visualizing information about the Earth's surface over the last 40 years. Geographical Information Systems (GIS) provide the functionality to perform these tasks and are now the preferred tool of many organizations within the administrative, commercial, governmental and voluntary sectors. This type of software is also becoming important within education not only for the particular subject areas traditionally associated with managing the features and understanding the processes operating at the Earth's surface, but also in those areas concerned with describing the world around us.

Issues relating to the data that GIS require in order to function has been a significant, but sometimes underplayed topic during the development of the technology. Not surprisingly, the conversion of existing geographical data from an analogue, usually paper-based, medium into a digital format was seen as the main problem and consumed many person days in GIS projects during the early stages of the system's development. There is now a proliferation of geographical data available and the cost of acquisition has reduced significantly, so much so that suppliers of GIS software often provide a number of datasets at least nominally free of charge to their customers. Geographical data can be divided into two broad categories – spatial data about the geometry, location, shape and relationships of geographical phenomena and attribute or variable data describing or quantifying these features.

There has been a remarkable growth in the volume of geographical data in both categories, especially in the amount captured directly into a digital format, thus removing the need for end-users to undertake the time-consuming process of conversion. These data are becoming more accessible to users of GIS as well as some of the more commonplace information processing technologies as the internet has opened up the opportunity for accessing and transferring data across the globe. The development of internet-based GIS has also obviated the need for users to acquire the data themselves, since these systems deliver some spatial data handling functionality over the World Wide Web rather than from software operating from the analyst's desktop.

A number of books at different levels of complexity have been published that seek to explain how GIS works and to demonstrate applications of the technology in different arenas. This book concentrates on the characteristics and sources of geographical data that can be used in GIS. It is intended to act as a guide for those people already familiar with GIS technology, but who may contemplate branching out into new applications areas and to novice users who are untutored in the nature of geographical data as well as the software used for handling this information. Part 1 examines the generic issues of geographical data in terms of access and looks at how data may be classified. The four substantive chapters in Part 2 focus on particular categories of geographical data – remotely sensed, cartographic, censuses and surveys and administrative and statutory data. The characteristics of these types of data are examined by reference to examples from the UK and other countries. It would be impossible and largely futile to outline the availability of digital geographical data for small spatial units across all countries. Instead a series of countries has been chosen from Asia, Australasia, Europe and North America in order to provide some contrasts and to indicate the direction in which the provision of geographical data is moving. Inevitably more countries were left out than could be included in the space available and omission is not intended to imply lagging development in digital geographical data provision. Part 3 illustrates some of the problems that may be encountered when undertaking analysis of geographical data to compare different places, to examine change over time or to integrate information from different sources. Finally, continuing developments are considered with reference to the internet-based GIS, online access and potential new sources of geographical data.

Nigel Walford
Wivenhoe, Essex

Acknowledgements

It seems as though geographical data have been central to my thoughts for a number of years, at least since the Data Archive brought me into contact with digital population census statistics and other kinds of data and with Geographical Information Systems. I am grateful to former colleagues there and to current colleagues at Kingston University for enabling me to continue my interest in this changing field. Geographical data are a constantly renewable resource and what seemed like innovative means of access and applications only a few years ago have now been overtaken by events. I am grateful for the assistance given to me by various people in writing this, not least Colin Barr at the Centre of Ecology and Hydrology, Ray Bollman in Statistics Canada, Chris Denham in Census Division of National Statistics, Chris Templeton in the Department of the Environment, Food and Rural Affairs and Steve Bisco of Colchester Borough Council.

Some of the tables and figures include Crown Copyright material made available through the Data Archive, University of Essex, Manchester Information and Associated Services (MIMAS), University of Manchester and Edinburgh Information and Access (EDINA), University of Edinburgh. I would also like to acknowledge the various internet websites that have been useful sources for the summaries of some of the datasets that have been discussed. The tables in Chapter 8 make reference to most of these sources, although some have been mentioned elsewhere in the book.

There have been two people who have played key roles in helping me to see this book through to its conclusion. Claire Ivision, Cartographer in the School of Earth Sciences and Geography, Kingston University has worked tirelessly to produce the figures. She has responded with calm equanimity when asked to make changes, let alone to do just another new figure with only 10 days to go before the publisher's deadline and corrections still to be made on others. My wife, Ann Hockey, has been tireless and, more importantly, constantly enthusiastic and helpful, in her efforts to keep me on track with completing this book project. To these, and no doubt others, I owe a considerable debt of gratitude.

Nigel Walford
School of Earth Sciences and Geography
Kingston University

Abbreviations List

ABS	Australian Bureau of Statistics
ACORN	A Classification of Residential Neighbourhoods
ACG	Address Coding Guide (USA)
ANZLIC	Australia New Zealand Land Information Council
AuSLIG	Australia Land Information Group
AVHRR	Advanced Very High Resolution Radiometer
CAS	Census Area Statistics (UK)
CEH	Centre for Ecology and Hydrology
CGIS	Canadian Geographic Information System
CORINE	Co-ordinated Information on the European Environment
CPH	County, Parish Holding (number)
CSV	Comma Separated Variable
CTIS	Centre for Topographic Information at Sherbrooke (Canada)
DBMS	DataBase Management System
DEFRA	Department of the Environment, Food and Rural Affairs (UK)
DEM	Digital Elevation Model
DETLR	Department of Transport, Local Government and the Regions (UK)
DETR	Department of the Environment, Transport and the Regions (now partly DEFRA and DETLR) (UK)
DLG	Digital Line Graph (USA)
DOQ	Digital Orthophoto Quadrangle (USA)
DSS	Department of Social Security (now Department of Work and Pensions) (UK)
DXF	Data eXchange Format
ED	Enumeration District (UK)
EDINA	Edinburgh Information and Access
EMR	Electromagnetic Radiation
EROS	Earth Resources Observation System (USA)

ESRC	Economic and Social Research Council
FADN	Farm Accountancy Data Network (Europe)
FBS	Farm Business Survey (UK)
FIPS PUBs	Federal Information Processing Standards Publications (USA)
GDF – DIME	Geographic Data File – Dual Independent Map Encoding
GHS	General Household Survey (UK)
GIS	Geographic(al) Information System
GLIS	Global Land Information System (USA)
GPS	Global Positioning System
GROS	General Register Office for Scotland
GSI	Geographical Survey Institute (Japan)
GSI	Geological Survey of India
HTML	Hyper Text Macro Language
HTTP	Hyper Text Transfer Protocol
ICT	Information and Communications Technology
IFEN	*Institut Français de l'Environment* (France)
IFOV	Instantaneous Field Of View
IGN	*Institut Géografique National* (France)
INSEE	*Institut National de la Statistique et des Études Économiques* (France)
IRS	Indian Remote Sensing
ISRO	Indian Space Research Organization
JODC	Japan Oceanographic Data Centre
JUVOS	Joint Unemployment and Vacancy Office Statistics (UK)
LBS	Local Base Statistics (UK)
LINZ	Land Information New Zealand
LIS	Land Information System
MAFF	Ministry of Agriculture, Fisheries and Food (now DEFRA)
MEGRIN	Multipurpose European Ground Related Information Network
MIMAS	Manchester Information and Associated Services
NAPP	National Aerial Photographic Programme (USA)
NCGI	*Nationaal Clearinghouse Geo-Informatie* (The Netherlands)
NERC	Natural Environment Research Council
NGDF	National Geospatial Data Framework (UK)
NHSCR	National Health Service Central Register (UK)
NISRA	Northern Ireland Statistical and Research Agency
NLUD	National Land Use Database (UK)
NOAA	National Oceanic Atmospheric Administration
NOMIS	National Online Manpower Information Service (UK)
NRSA	National Remote Sensing Agency (India)
NSDF	National Spatial Data Framework (USA)
NSS	Neighbourhood Statistics Service (UK)
NTF	National Transfer Format
NUTS	*une Nomenclature des Unités Territoriales Statistique* (European Union)
OA	Output Area (UK)

ONC	One Number Census
ONS	Office for National Statistics (now NS National Statistics) (UK)
OS	Ordnance Survey (UK)
OSNI	Ordnance Survey for Northern Ireland
RSAS	Reformatted Small Area Statistics (Scotland)
SAS	Small Area Statistics (UK)
SDI	Spatial Data Infrastructure (Australia and New Zealand)
SNIG	*Sistema Nacional de Informação Geográfica* (Portugal)
SNZ	Statistics New Zealand
SoI	Survey of India
SPOT	*Système Pour l'Observation de la Terre* (France)
STF	Summary Tape File (USA)
TIGER	Topologically Integrated Geographical Encoding and Referencing (USA)
TIROS	Television InfraRed Observation Satellite
UPRN	Unique Property Reference Number
USCB	United States Census Bureau
USGS	United States Geological Survey

PART 1
GEOGRAPHICAL DATA

Chapter 1

Geographical Phenomena and Geographical Data

1.1 CLASSIFICATIONS OF GEOGRAPHICAL DATA

The development of spatial or geographical awareness can be traced back to the earliest civilizations, when people wanted to know where things were and to record information about the world around them. Such societies were interested in where particular features in the landscape were to be found, to distinguish between places that were suitable or unsuitable for different purposes, and to know where they were located in relation to these features and to other tribes or social groups. In other words they were interested in how much of what is where. Although most societies in the world have clearly moved on a long way from these more primitive times, the basic concerns of modern societies' spatial questions are much the same. These questions are still fundamental in enabling people to navigate between different places, to organize their lives and to utilize the resources around them. In order to answer these questions we need to assimilate data about the world into information that can be understood and communicated from one person or organization to another. Further, although current information is usually of primary importance, there are occasions when intergenerational communication is also relevant.

Before looking further at these issues one fundamental question has to be addressed. What are geographical data? Probably the best way of answering this question is by going back to 'first principles'. Data, or datum in the singular, is derived from Latin and means having been, or that which is, 'given'. A similar concept is contained in the French translation of data as *données*, which also means 'given'. In other words, data are those things on which understanding and explanation are based. Put more simply, data are given facts about phenomena. For example, one fact attached to the phenomenon known as pure water is that it boils at 100°C. This characteristic enables a scientist, who is presented with an unknown liquid that also boils when heated to 100°C, to deduce that it must be water, unless some other liquid also possesses this trait. If geography is concerned with the description and explanation of things occurring on or near the Earth's surface, then geographical data are facts relating to features which are spatially referenced to this surface. Again put more simply, geographical data are facts about the world around us. However, there is one further complicating factor, which will be examined in more detail later, and that concerns the nature of the features or phenomena to which the geographical data relate. Suffice to say at present that the Earth's surface is composed of a mass of interacting and interconnected features, some of which possess fixed and others mutable locations.

A number of different methods for classifying such geographical facts have been devised to help with understanding the nature of the various types of geographical data. These classifications are attempts to bring some order to the potentially confusing array of data that are available. One classification refers to the relationship between the user and collector of the data and distinguishes between primary and secondary sources. This primary versus secondary division refers to the source of the data and not to the data themselves, and is not unique to geographical data. A primary source of geographical data is one in which the person or organization responsible for data collection also uses and analyses them. A secondary source, on the other hand, implies that the person or organization that is currently using or analysing the data was not involved with their collection. Such sources are sometimes referred to as archival sources (Shaw and Wheeler, 1985; Matthews and Foster, 1989).

The importance of this division lies in the level of influence that the user of the data has over the precise nature of the dataset. The points at issue relate to how concepts and information are defined, the number of observations in the dataset, the precise area covered, the frequency of updating the information, and the way the data are structured for computer-based management and analysis. In the case of a primary source of data, the analyst or investigator has greater control over these matters, but with secondary sources one or more of these crucial characteristics has been determined by the data collector. If a dataset containing demographic information is used in an examination of population change in rural settlements, it is important to know how settlements are delimited. A number of possibilities present themselves to the person or organization collecting the data, ranging from using administrative areas through to defining a boundary around each village and hamlet. The first option might result in several individual settlements being lumped together, whereas the latter might lead to the omission of outlying dwellings and farms or their attachment to the nearest settlements. The underlying issue relates to the problem of defining the unit of observation in geographical data.

Regardless of whether geographical data are obtained from a primary or secondary source, the method by which the data are collected and presented provides another basis for classification. Essentially four main categories of quantitative data are identified:

- remotely sensed
- cartographic
- censuses and surveys
- administrative and statutory

The distinctive characteristic of the means by which remotely sensed data are collected is that there is no direct contact or human intervention with the phenomena being measured. In other words there is some type of measuring instrument between the phenomena and the observer. Remotely sensed data are derived from three main sources: aerial photographs, satellite imagery and laboratory equipment. Cartographic data refers to information obtained from existing maps, which are themselves the result of processing and abstracting from data obtained by direct field survey and/or remote sensing. Administrative and statutory data are sources that are the by-product of some administrative process or statutory requirement, which are useful for geographical analysis. Statistics produced from the process of birth and death registration exemplify this geographical data source. Censuses and surveys are distinguished from administrative data by virtue of the fact that the purpose of the data collection exercise is to generate information for answering research questions or information that is not routinely required.

Non-quantitative sources of information, such as oral histories, pictures and photographs, are also used in geographical analyses. Contemporary information technology permits the rendering of these sources into a machine readable form, which occurred in the case of the BBC Domesday Project in the mid-1980s (Openshaw *et al*, 1986). However, the inherently qualitative character of this type of information means these sources lie outside the main focus of this text. The four types of geographical data source identified above are examined in detail in subsequent chapters.

A further important way of classifying geographical data relates to the contents or substantive component of the dataset. Given that geographical data are viewed as data in which the entities or phenomena are spatially referenced to the Earth's surface, it follows that datasets covering a very wide range of topics fall within this definition. One basic substantive division relates to the human–physical split within academic geography, although even this is not watertight in view of cross-cutting environmental topics, in which interaction between people and the physical environment is a primary concern. Nevertheless, the human–physical split is a useful one, which can clearly be broken down further into a number of sub-topics. The problem with any substantive categorization of data, whether or not they are geographical, is that the specific topics contained in a dataset may be relevant to a number of different areas of academic or professional endeavour. Information relating to housing tenure may be of interest to the housing economist as well as those concerned with the social characteristics of the tenants. In a similar vein, information on soils is relevant to agriculturalists as well as foresters. Any substantive or topic-orientated classification system must take into account the overlapping purposes for which a particular attribute or variable may be used.

1.2 GEOGRAPHICAL DATA ENTITIES

Given the definition of geographical data as data concerning features which are spatially referenced to the Earth's surface, this does not necessarily imply that the phenomena are inherently spatial themselves. For example, one set of geographical data might relate to phenomena that can be spatially fixed to a particular location on the Earth's surface at a certain time, but that on another occasion some or all of these phenomena may have moved to a different location. In the case of a dataset containing information about the recreational activities and socio-economic characteristics of a group of people who are using a national park, the geographical component of the data refers to the 'fact' that they undertook this activity in the particular national park and not elsewhere. In the case of a dataset covering facts about the chemistry of water in a particular reach of a river, the facts will be about the chemical molecules present at the time when the water was measured. Since rivers do not contain water which is 100% pure, any samples analysed on other occasions would be different from those taken previously. Water is by its very nature a fluid entity and is only constituted as a river by virtue of being contained between two earthen or rocky banks.

In both these cases the geographical component of the data is changeable, because the phenomena being measured or examined are capable of altering their spatial location. The people in the national park will leave to return home, to go to work, to visit friends or to undertake one of a myriad of other activities. This change of state means that the individuals not only change their location but also cease to be recreationists and take on some other socio-economic relationship, such as employees or household residents. The water in the river will also change its spatial location and the combination of molecules present in a given volume of water will alter. The change of location and of state in the case of the visitors to the national park may be viewed as the result of voluntary or purposive action on their part, whereas in the aquatic case the changes are of course involuntary and determined by physical processes.

In other cases the phenomena contained or represented in the geographical dataset are themselves spatial in character. The information in such datasets relates to phenomena whose spatial reference tends to be rather more permanent, although even in these instances locational variation might occur but over a longer time span. The manner in which spatial features are described has been referred to as 'formal spatial languages in geography' (Harvey, 1969: 212). Throughout the development of academic geography the prevalent spatial language has been that of Cartesian geometry, which relates to the notion of absolute space. However, within recent decades the arguments in favour of a relative view of space have received greater attention. The debate between the supporters of each scheme, which experienced a lull after the fervour of the spatial revolution, has been brought back into sharper focus by the growing of interest in geographical information systems.

Inherently spatial entities are usually divided into three data primitives, points, lines and areas (Star and Estes, 1990; Burrough and McDowell, 1998), with the further complication of surfaces or volumes in three-dimensional space. A point feature has no linear or areal extent and simply occupies a single location. A linear feature represents the connective path between at least two points. An area or polygon is a feature which is defined and contained by an encompassing boundary line or lines. Many types of polygonal feature do not exist in isolated individual instances, but are collections of conceptually similar areal units, which cover a region, country, continent or the Earth. A surface feature may be defined as the three-dimensional representation of a point, line or area. For example, an oil well, a point feature when depicted in two dimensions, becomes a vertical column when shown three dimensionally. Similarly the line of a river becomes a channel and the area of a lake becomes a water body.

Geographers have debated long and hard over the spatial unit which is most suitable for their studies. Up to the 1950s, many people regarded the region as the natural unit for the study of geographical information. These early geographies were therefore an assemblage of information on the human and physical aspects contained in a region reflecting the absolute view of space. Unfortunately such regions are difficult to define or delimit, since the social, economic and cultural variables being examined do not exhibit abrupt changes from one region to another, which is implied when a 'solid' boundary line is drawn. In fact the values for most variables shade into each other, and even if some variables do have sudden changes, these might not occur at the same locations in each case. For example, the limit of the built-up area provides one reasonably clear boundary line for the physical extent of a city. However, the boundary of the city in terms of employment and retailing may extend into the countryside beyond. Thus, in respect of these three functional characteristics, the city may have completely different boundary lines.

Apart from these definitional difficulties, there were two other reasons why regions fell from favour. Even if a region could be satisfactorily defined, it remains a unique entity or phenomenon. The Basque region in south-west France is unique, although it may possess similar characteristics to other regions of Europe whose inhabitants also feel that their regional identity is being lost in a general socio-economic levelling process. Regions that are not defined on the basis of the cultural perceptions of the residents, but with respect to administrative convenience, such as the government office regions of the United Kingdom and the *Nomenclature des Unités Territoriales Statistique* (NUTS) statistical units of the European Union, are also composed of unique assemblages of characteristics. The uniqueness of each region means that although they can be compared and contrasted in terms of their qualitative and quantitative features, it is not possible to derive any universally applicable laws pertaining to such phenomena. In the 1950s and 1960s geographers became much more interested in attempting to specify scientific laws and as a consequence many rejected the region as the natural unit for geographical investigation. The focus switched to the search for laws of spatial location by which physical, social, economic and political processes generated reproducible spatial patterns.

The move towards smaller spatial units provides the third reason for the declining importance of regions as geographical entities. Part of the explanation for this trend is attributable to the ease with which it is now possible to store information in an accessible and easily assimilated form in a physically smaller space that was previously possible. For example, the full set of Small Area Statistics from the 1991 UK Population Census, representing some 8000 individual counts, for each of 130 000 enumeration districts, is available on a CD-ROM disk. The equivalent amount of storage in the era of punched card computer storage is estimated as some 130 million punched cards and in the pre-computer era 3240 volumes of statistical tables occupying 130 m of library shelving. However, it would be wrong to associate the move towards 'small is beautiful' in terms of geographical units solely with technological developments in information technology.

There are equally persuasive theoretical arguments that urge the investigator of geographical problems to favour smaller rather than larger spatial units. The change occurred because a range of physical, social, economic and political processes are believed to operate on individual or identifiable spatial features such as drainage basins, farms, households, glaciers, roads and settlements. It was these classes of phenomena that the geographers should become concerned with rather than the assemblages of diverse features found within a region. During the process of transition there were attempts to identify regions on a more systematic basis, such as agricultural or farm-type regions, but even these have now receded into the background as greater emphasis on diversity and a multiplicity of explanation has surfaced with the emergence of post-modernism, which is itself compatible with the use of smaller spatial entities.

The combination of these theoretical developments and technological changes means that there is considerable variety in the range of entities represented in geographical data. The sources of geographical data covered in this book will reveal many different types of entity, but it cannot hope to be complete and exhaustive. At this point it is worth summarizing how these entities might be classified (Table 1.1). The typology has two main dimensions. The spatio-temporal permanence of the entity, how long it exists in a given state, forms one dimension, and its linear or areal size while existing in that state, provides the other.

1.3 GEOGRAPHICAL ATTRIBUTES AND VARIABLES

The spatial dimension is not the only quantifiable characteristic of geographical entities. In many datasets the phenomena are also measured in respect of different variables and distinguished from each other by qualitative attributes. Thus, a collection of attributes and/or variables is usually associated with each entity. It is important not to confuse the term qualitative in the context of statistical analysis with its use in the methodological sense, when it refers to ethnographic and humanistic approaches to geographical investigation. The latter is in some respects antithetical to the largely numerical (quantitative) datasets discussed in this text and has emerged as a key dimension in late twentieth-century human geography. Qualitative geographical data may quite feasibly be quantitatively analysed through such techniques as

Table 1.1 Examples of geographical (spatial) entities classified in terms of space and time scales

	Spatio-temporal permanence measured in:									
Space	Seconds	Minutes	Hours	Days	Weeks	Months	Years	Decades	Centuries	Millennia
μm or μm^2			Clay particle							
				Sand grain						
	Molecule									
mm or mm^2							Stone			
cm or cm^2	Water temperature					Spring				
m or m^2							Household			
								Dwelling		
								Field		
							Oil well			
dm or dm^2							Postcode			
								Factory		
								Lake		
km or km^2		Commuter journey				Ocean wave				
							County			
								National Park		
									River	
								Coastline		
										Sea
										Mountain range

Note: Spatial and temporal scales are indicative only.

categorical data analysis (Wrigley, 1985). Nevertheless, there is a greater difference between those forms of research which employ oral and textual information sources and those containing numerical or alpha-numerical attributes and variables.

Qualitative data in this respect denotes the attributed, as opposed to the measured, characteristics of geographical entities, such as type of vegetation cover, occupational group, household tenure, climatic zone and settlement type. These attributes are inherently descriptive in their nature and each particular individual in a class of geographical phenomena belongs to a particular category. Thus household tenure can be exhaustively described and categorized as owner occupation – purchasing with a mortgage or owned outright, public sector tenancy and private sector tenancy – furnished or unfurnished. Households can be separated from one another in respect of the category to which they belong. These attributes enable counts of the occurrence of entities within the different categories to be made. Thus aggregated data may be obtained, yielding, for instance, the number of households living as private unfurnished tenants within a given spatial entity. These counts are therefore quantitative values for a qualitative characteristic.

In most of the datasets used in geographical information systems each individual entity has its own identification label, often known a unique reference number or unique property reference number in the case of gazetteers and land information systems. The purpose of such attributes is to ensure that occurrences of the spatial feature are not confused in the database or by the analyst, and to query the data about the characteristics of a particular entity. In addition to substantive and labelling attributes, many geographical datasets also contain variables that store numerical measurements in respect of the entities. These quantitative measures differentiate between the entities in respect of some order of magnitude. In the example referred to previously, some visitors to the national park may have travelled over 15.0 km, others between 10.0 and 14.9 km and the remainder less than 10.0 km. Visitors can therefore be differentiated sequentially, but imprecisely, in terms of the distance travelled. Such variables are referred to as measured on the ordinal scale. If the distance of each visitor's journey had been accurately enumerated to the nearest kilometre and hence measured on the interval scale, it would be possible to conclude that someone coming 8.0 km had travelled twice the distance of another person starting out 4.0 km away. Suppose also that the survey included respondents who had previously visited the park and they were asked to recall the distance of their journey on that occasion. The difference between the measurements could be calculated to produce a variable measured on the ratio scale.

The distinction between the interval and ratio scales essentially relates to two factors: whether the measurement units have an absolute or arbitrary zero; and whether two pairs of values measuring the difference between entities have identical or different ratios. Interval scale variables contain exact numerical values so that the difference between any two entities can be measured, but when different scale units are used (e.g. Fahrenheit and Celsius for temperature) the difference between any two pairs of measurements will not have identical ratios. On the other hand, the distance between two locations measured using different scales (e.g. metric and imperial) will have an identical ratio to each other. Distance between places has an absolute zero since any two locations cannot be a negative distance apart. On the other hand the difference in distance between the two journeys in the national park example might have a negative value, indicating that one journey was shorter than the other. The nature and implications of the differences between the ordinal, interval and ratio scales of measurement, and between attributes and variables, are fully elaborated in most statistics texts (see for example Silk, 1979: 6–8; Walford, 1995: 13–16). For the present purpose it is sufficient to note that these differences exist and that adequate details about the characteristics of the attributes and variables held in a dataset should be available with the information provided about secondary sources of geographical data.

1.4 SPATIAL AND TEMPORAL DATA REFERENCING

Spatial and temporal referencing of data was first summarized diagrammatically in the form of the 'geographical data cube' (Harvey, 1969; Shaw and Wheeler,

1985). This conceptual model, although initially devised to represent the multidimensional nature of geographical investigations, may be considered as a framework for positioning geographical datasets in respect of three classificatory dimensions – space, time and substantive content or characteristics. Figure 1.1 illustrates how point, line and area phenomena can change their shape and their attribute/variable values over time. The settlement in the top right has progressively increased in area and population, the fields of the farm have been amalgamated, the river has altered its course and flow rate, and the number of churches has fluctuated with the growth and decline in the number of parishioners. The purpose of this section is to look at the practical issues associated with the inclusion of spatio-temporal coding or referencing systems in geographical datasets. Such systems are fundamentally concerned with placing or, quite literally, locating entities contained within geographical datasets into a spatio-temporal framework.

1.4.1 Spatial Data Referencing

The form and nature of the spatial or locational references of geographical phenomena are of crucial importance in the analysis, integration and manipulation of geographical datasets. One of the distinguishing characteristics of these datasets is that, not only are the attributes and variables with respect to which the geographical entities are measured capable of analysis, but the geometrical properties of the features themselves may also be manipulated by adjustment to their spatial references. For example, in the unreal, virtual world of the computerized geographical database, a factory chimney stack issuing pollutant emissions can be repositioned prior to any such change happening in the real world. Furthermore the effect of making such a change on the environment can be modelled within the confines of the database in order to investigate a variety of new locations for the chimney. This simple example illustrates one key characteristic of geographical datasets, namely that the location and properties of the spatial features which are represented can be altered.

There are a number of different ways in which spatial references are specified in datasets and in some cases more than one might be used. When examining the nature of spatial referencing systems there are two questions to be considered. How is the geographical entity's location referenced to the Earth's surface? How is this spatial reference represented in a digital form in the dataset? There are two main ways in which a geographical entity is referenced to the Earth's surface. One method involves some form of alpha-numerical index, including simply the naming of places, while the second is a system of grid references. The digital representation of spatial references to the geographical entities under these systems divides into three main types:

- embedded within the data file
- linked to the entities via a key pointer
- implied by the structure of the data model

Figure 1.2 exemplifies the classification of spatial referencing systems in geographical

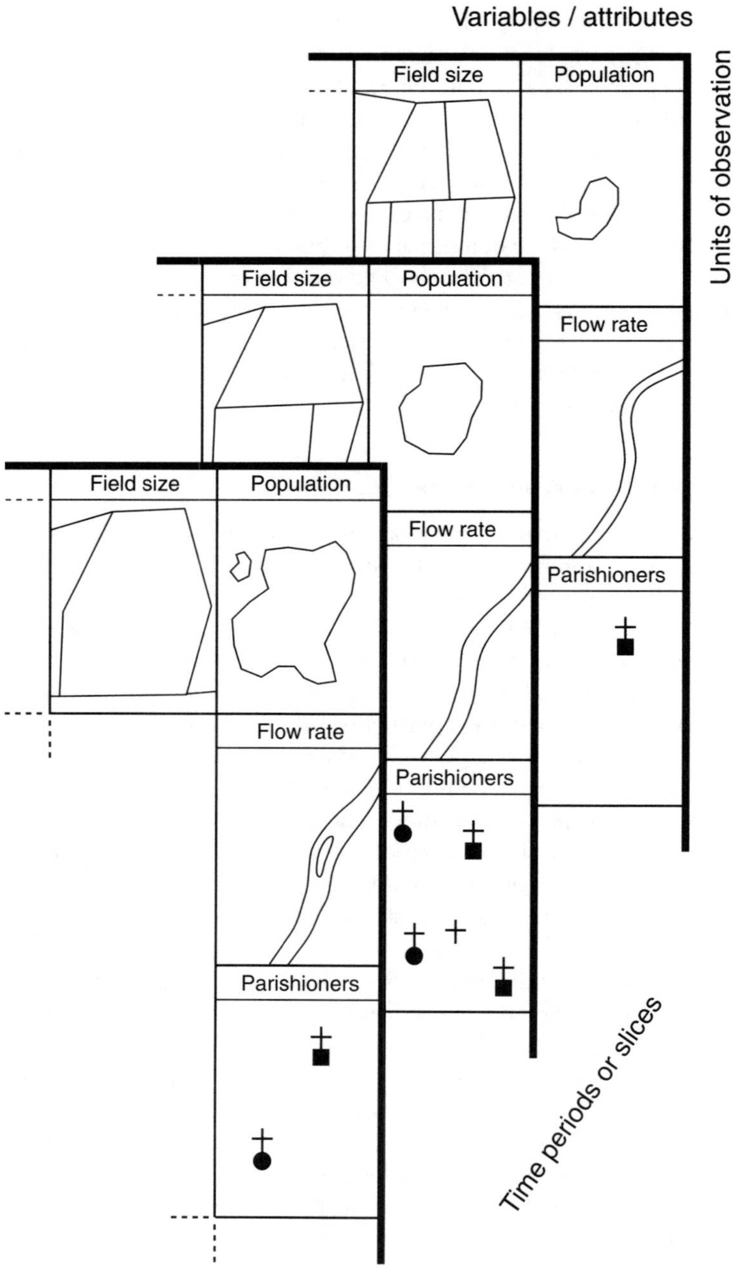

Figure 1.1 Spatio-temporal change in geographical entities.

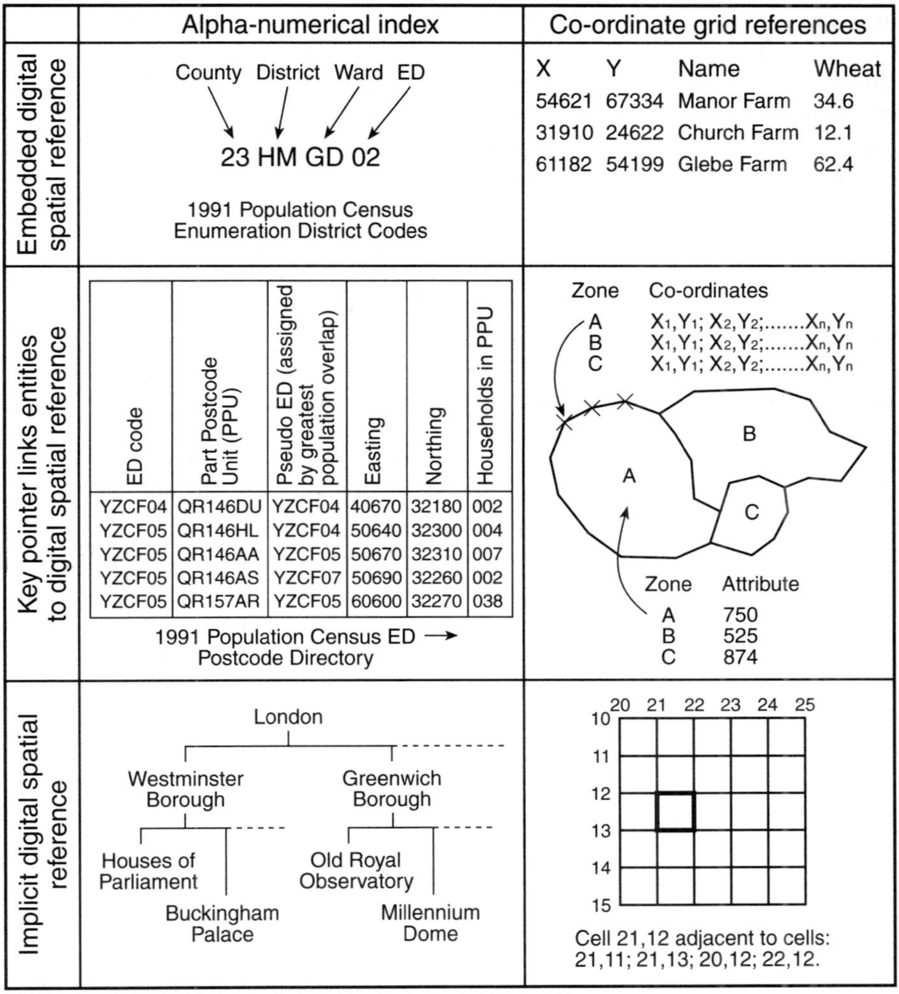

Figure 1.2 Classification of geographical datasets according to the georeferencing system.

datasets according to these two dimensions and provides point of reference for examining the spatial characteristics of specific datasets in later chapters.

1.4.1.1 *Alpha-numerical Indices*

Alpha-numerical indices exist in different forms as a means of referencing various types of geographical feature. The spatial entities to which these indices relate often have a nested hierarchical form, as in the case of administrative areas. For example, in France, *communes* lie within *départements*, which in turn combine to form *regions*. Similar structures exist in many countries, although interestingly the nationwide hierarchy in the United Kingdom (districts/boroughs, counties and regions) is being dismantled in favour of a mixed system at the lower two tiers. These hierarchies, although common in the case of areal features, also occur with respect to point and line features and can even cut across feature types. For example, first, second and third order streams combine to form a drainage system or catchment area.

Perhaps the simplest form of spatial reference is the place name, which, although imprecise in comparison with grid co-ordinate systems, nonetheless identifies the location of a feature. The place named Kingston, for example, is found in Ontario, Canada, and in Jamaica; in addition there are others with a suffix such as Kingston on Hull and Kingston upon Thames in the UK. In a similar vein, there can be few cities or towns in the United Kingdom without a road named 'High Street'. Likewise, multiple instances of point features occur, for example public houses called the 'Red Lion' and the 'King's Head', to say nothing of restaurants called 'McDonalds'. These examples show that in many cases a place can only be uniquely identified and located by a combination of names, such as 'Boston, Massachusetts', 'High Street, Ashford, Kent', 'Red Lion, Bletchingly, Surrey' or 'McDonalds, Cowdrey Avenue, Colchester, Essex'. The imprecision of such names means that they are not very useful for analysing geographical datasets, other than for use as gazetteers when searching for places.

A more useful type of index is one which provides a unique code for each individual occurrence of a particular type of feature. These codes can be composed of either letters, numbers or a combination of both (Figure 1.2). These codes are either simple or complex. A simple code may comprise a numerical sequence, whereby numbers are assigned to entities in an unstructured way. One example of a simple code might be a unique reference number used for towns sampled in a survey. However, simple codes more commonly occur in datasets covering parts of a country or region. For instance, continuing with an earlier example, a survey of visitors to several national parks might simply distinguish between each by an alpha-numerical code. Complex codes usually incorporate a hierarchical structure, in which each level relates to a distinct type of geographical feature, which is often named. These complex codes thus have the key benefit that a specific level in the hierarchy can be selected and all the constituent features occurring at that level will be included. The hierarchy will often relate to a large area, such as a country, which can be divided into these different levels. Examples of complex codes applicable across the UK are illustrated in Figure 1.3.

Raper *et al* (1993) have discussed in detail the origins and nature of the postal geography of the United Kingdom and reviewed the general characteristics of such systems in other countries. The UK system was introduced by the Royal Mail in 1971 and for most of the country uses a seven-digit, four-level hierarchical code to specify postal areas (Figure 1.3a). The two initial letters of a postcode denote the postal town and the following one- or two-digit number relates to a postal district served by that town. These two elements are known as the inward part of the postcode. The outward part of the code is also composed of two elements: a single digit number defines the postal sector and finally two letters specify the unit postcode. The reference to the inward and outward parts of the postcode points to the relationship with a postal sorting office and is a reminder of the origins of the postal referencing system. Postal delivery points with in excess of 25 items of mail per working day in urban areas (50 items in rural areas) are assigned a large user postcode, which has the same structure as the standard postcode unit but only relates to the one address. Most countries have adopted some form of postal referencing system, although many are not as spatially specific as the one found in the UK.

The usefulness of the UK's postcode referencing system for non-postal purposes was starting to emerge before the publication of the Chorley Report (HMSO, 1987), and has become more significant in subsequent years, partly in response to the recommendation that 'the preferred bases for holding and/or releasing socio-economic data should be addresses and unit post codes' (HMSO, 1987: 89). The Chorley Committee into the *Handling of Geographic Information*, recognizing that a unit postcode is not 'truly a spatial unit unless its boundaries have been defined' (HMSO, 1987: 88), based this recommendation on their perceived advantages, namely that they:

- are sufficiently small to facilitate aggregations to larger areas and to locally defined areas of interest
- permit the linking of address related data whilst preserving confidentiality
- are part of an established working system that is kept up to date by the Post Office
- provide full coverage of the whole country

The diffusion of postcode referencing into many UK datasets will be revealed in later chapters, however, it is should be noted that many organizations within various sectors of the UK economy responded positively to the Chorley Committee's recommendation and have adopted postcodes as a key means of spatial coding and of linking together data from different sources.

The second British example of a hierarchically structured complex code has been used for the 1981 and 1991 Population Censuses. In this case the four-level, eight-digit code relates to counties, districts (including London Boroughs), wards and enumeration districts (EDs). The ED shown in Figure 1.3b (43QYFU16) is an indivudal spatial unit at the bottom of the hierarchy. Enumeration districts (EDs) have until the 1991 Population Census been the data collection unit and, like the wards, districts and counties, they covered the entire country without omission or overlap. This feature enabled the upward aggregation of census statistics from ED to ward, ward to district and district to county to be achieved with relative ease. The growing importance and usefulness of postcodes has however meant that the importance of the traditional ED has changed, firstly in Scotland and more recently in England and Wales (see Chapter 6).

The third example of a complex code relates to farms in the UK, which are registered as holdings with the Department of the Environment, Food and Rural Affairs (DEFRA). Under this system each holding is assigned a County/Parish/Holding (CPH) number. The three-part hierarchy includes a county number, parish number and holding number. Parish numbers are likely to be replicated in different counties and holding numbers in different parishes. Although similar in some respects

(a)

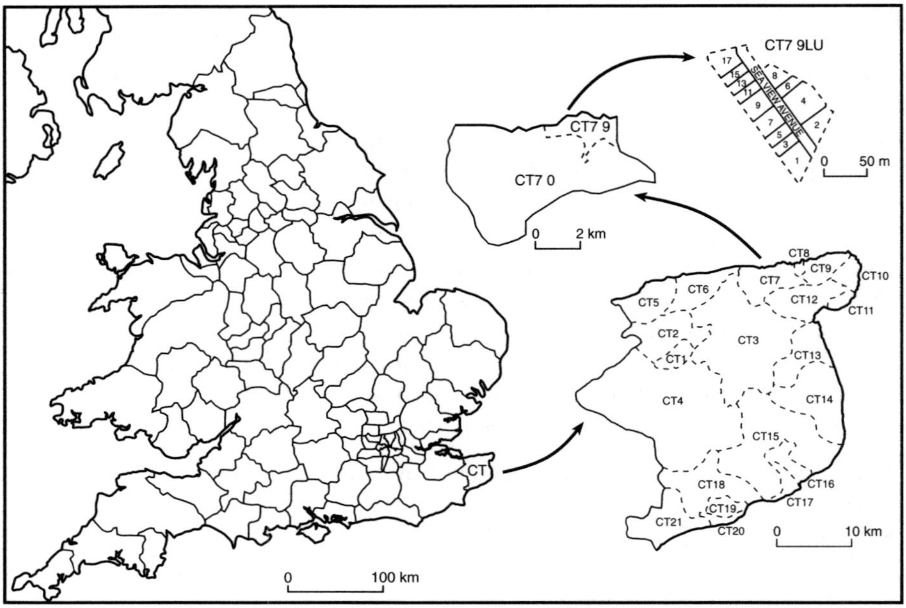

Figure 1.3 Complex hierarchical georeferencing systems for geographical data in the UK. (a) Part of UK postal geographical hierarchy, (b) administrative geographical hierarchy, (c) county/parish/holding numbers of British farms (hypothetical example).

to the postcode and census systems, CPH numbers differ to the extent that they relate only to agricultural land and not to the whole area of the country. A single farm may consist, from a business management point of view, of more than one holding and the spatial feature (i.e. the holding) referenced by the CPH is in some respects an area and in others a point feature. A further significant practical difference is that CPH numbers are maintained confidentially by MAFF, hence the example in Figure 1.3c is hypothetical, whereas the census and postcode systems are in the public domain.

1.4.1.2 *Grid Co-ordinate Systems*

The system of grid references was developed in the 18th century and was associated with the era of global exploration and the surveying of the Earth's surface. In many countries national grids have been devised which are tied in to the international longitude and latitude system, but which use a national reference datum point. For example, in Great Britain it was recommended that the Ordnance Survey, the national mapping agency, develop a National Grid in the 1930s. Ireland, north and south, although one of the British Isles, has its own grid system, which can be matched to that used in Great Britain.

A grid-based spatial referencing system works on the principle of superimposing a

(b)

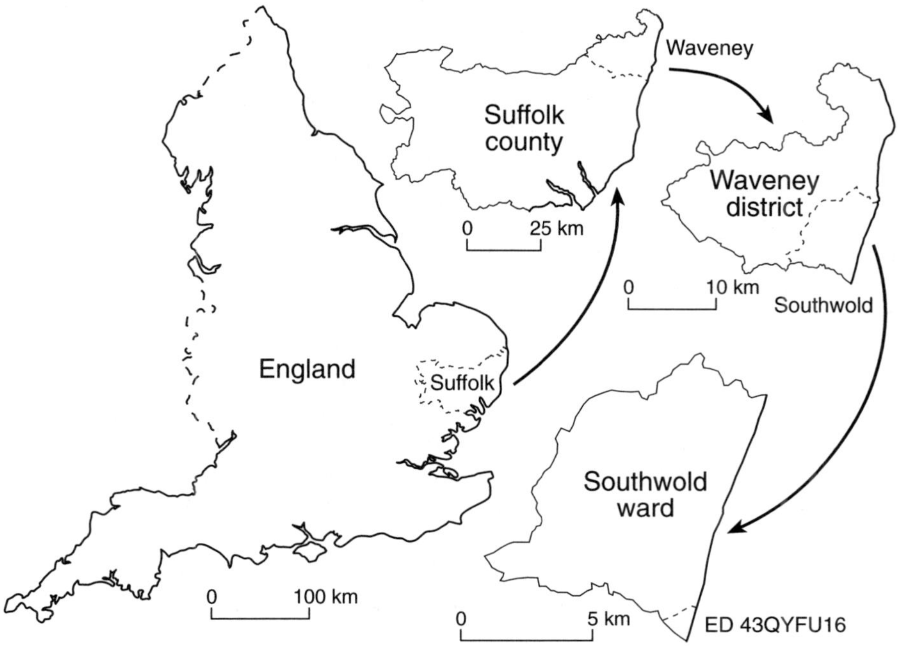
Waveney
Suffolk county
0 25 km
Waveney district
0 10 km
Southwold
England
Suffolk
0 100 km
Southwold ward
0 5 km
ED 43QYFU16

(c)

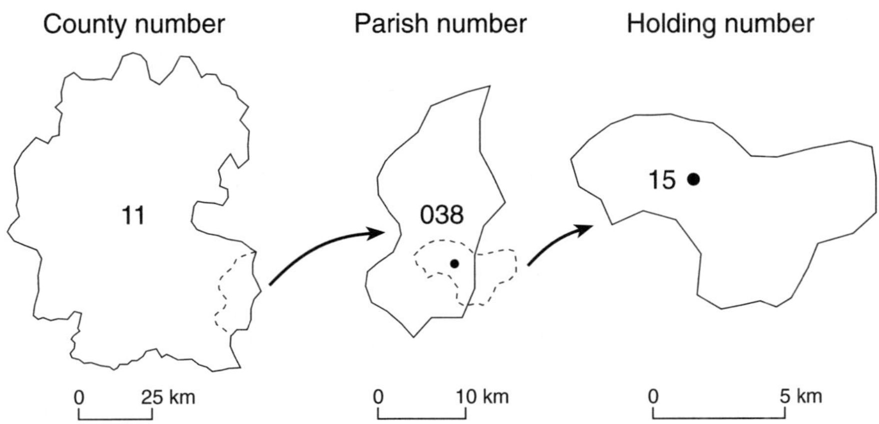
County number
Parish number
Holding number
11
038
15
0 25 km
0 10 km
0 5 km

regular grid, usually composed of squares, over the area of interest. Most grids are formed of vertical and horizontal lines numbered according to a numerical scale which enables distances on the ground to be related to those shown on a printed map or held in a computer database. This scale usually has a hierarchical structure enabling more or less precision to be employed in locating places on the Earth's surface. For example, grid lines which are scaled to 1 km can only locate features to this distance apart; smaller distance can be measured by interpolation. If the grid lines are 100 m apart then greater locational accuracy may be obtained. Another common feature of grid numbering systems is that there is an origin or starting point of zero. In other words, one horizontal and one vertical line are assigned the numerical value zero, and the origin of the grid occurs where these intersect (i.e. the grid reference is 0,0).

In the case of the grid composed of longitude and latitude lines which cover the globe, the numbering system uses degrees, minutes and seconds. The lines of longitude, which radiate outwards from and connect the North and South Poles, have their origin or zero line passing through the Greenwich Observatory in London, called the Greenwich (or Primary) Meridian. The zero latitude line is the equator and successive lines are numbered outwards to the North and South Poles. There are 360 lines of longitude (number of degrees in a circle), each of which is divided into 60 minutes and these in turn into 60 seconds. The lines of latitude are similarly divided, but in this case the degrees radiate out from the centre of the Earth between the right-angle formed by the intersection of one line through the North and South Poles and another along the plane of the equator. There are 180 lines of latitude, 90 in each direction outwards from the equator. Figure 1.4a illustrates how the system operates. Places on the Earth's surface are spatially referenced in the following manner: 30° 30′ 15″ Long., 0° 15′ 24″ Lat. North where ° denotes degrees, ′ minutes and ″ seconds.

National mapping agencies often replace the graticular network defined by the parallels and meridians of latitude and longitude with a regular square grid of Cartesian co-ordinates covering the relevant area of the Earth's surface. The vertical lines in national grids are often referred to as eastings (the line numbers become larger moving eastwards) and the horizontal lines as northings (the line numbers become larger moving northwards). The development of the National Grid in Britain paved the way for scaling on a metric basis, in place of the former imperial system, with the distance between any two adjacent vertical and horizontal lines capable of subdivision into 10 units. This process produces a new set of 100 smaller square units within each of the original squares. The process is illustrated in Figure 1.4b where the single large square forming part of 10 000 km^2 area labelled TM is divided into 100 smaller squares, and one of these has then been similarly subdivided. Places on the Earth's surface covered by the grid are spatially referenced in the following fashion: 382,161. The first three digits are the easting and the second group the northing. Each group can be thought of, using previous terminology, as a hierarchical numerical code. The three digits in each set locate the geographical feature in question to a square measuring 10 × 10 km within a 100 × 100 km square.

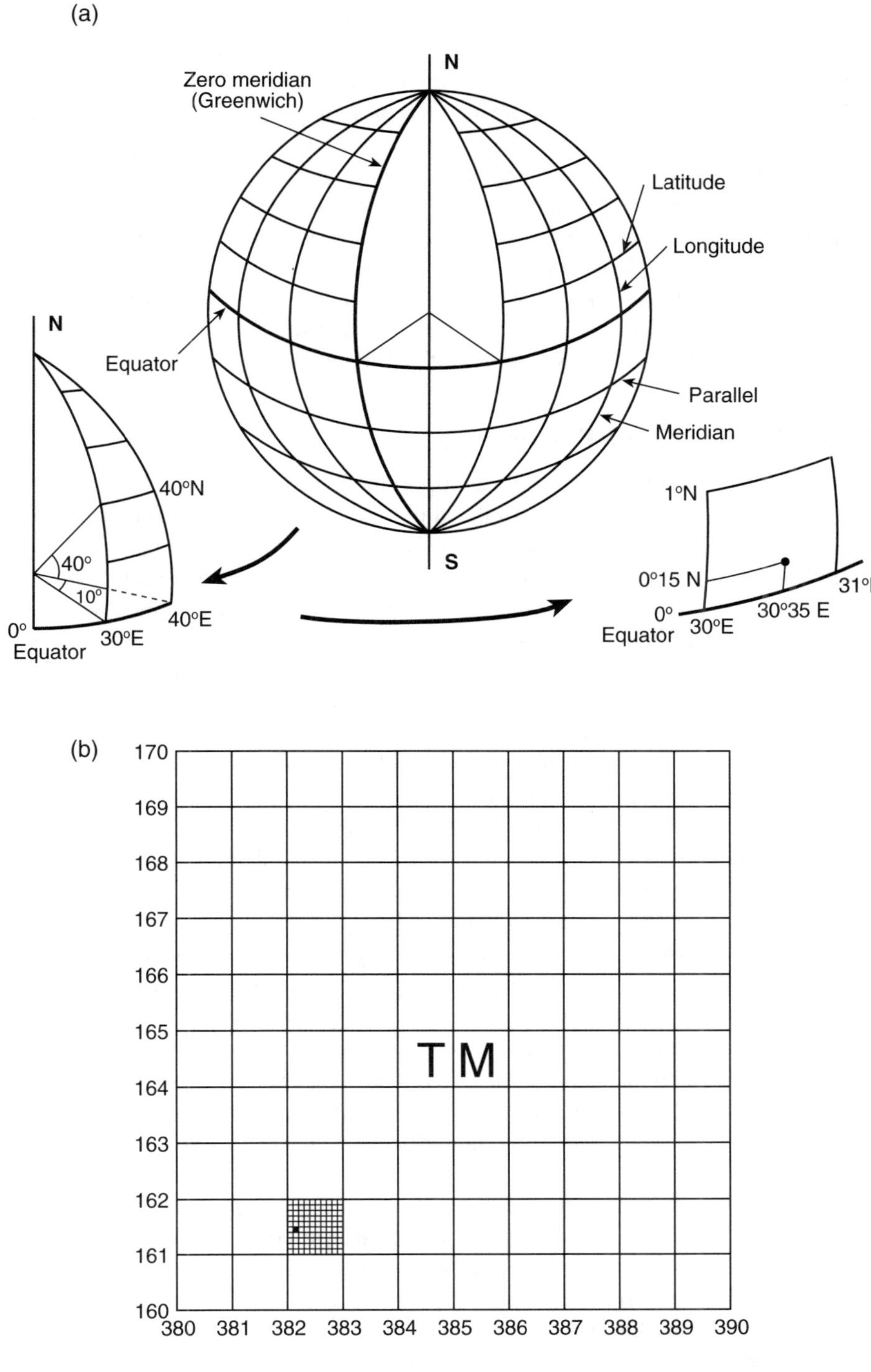

Figure 1.4 Global and national co-ordinate georeferencing systems. (a) Longitude and latitude, (b) British National Grid.

Grid referencing systems are commonly employed in street gazetteers and road atlases, for instance the A–Z Maps for most major cities and towns in Britain and the Automobile Association's 3 miles to the inch road atlas (Automobile Association, 1986). The grids used in such publications may or may not be linked into the longitude and latitude or a national system, which indicates the flexibility of such systems. One of the problems with grid systems is that although the public is becoming more familiar with their use for locating places on a map, people are not accustomed to knowing the grid reference for their residential address, place of work or any other location. Hence it is often necessary to use a central directory or look-up table to link an address or postcode with a grid referencing system in order to benefit from the greater level of flexibility and sophistication that is to be gained from the manipulation of a numerical system of spatial referencing.

1.4.2 Temporal Data Referencing

The temporal content and context are important features of geographical datasets. The temporal referencing of geographical data can also be undertaken at different levels of detail. In this case the underlying principle is that the time period or 'date stamp' of each phenomenon recorded in a dataset and each dataset itself are important variables. The former relates to the notion of temporal specificity and the latter to temporal frequency. Most data collection activities happen over an easily quantifiable period of time, whose length may range from an hour to a year or even longer. Although in some circumstances the duration of data collection may be only minutes or even a fraction of a second. For example automated weather stations may be synchronized to record simultaneously wind direction and other meteorological variables at different points on or over the Earth's surface. Therefore an important relationship exists between the length of time taken to collect a particular item of information and the period for which the obtained data value is valid. For example the temperature of a river may be measured over 30 s, 2 or 5 min periods in order to even out any supposedly minor fluctuations, but the value obtained for these three intervals may be significantly different. Similarly someone's employment position may be recorded in a survey as a full-time employee one day, yet the following week or even day, the person may retire or become unemployed, thus completely altering their economic and social position in society. In a survey lasting several weeks or even months, it is assumed that such changes of circumstance will balance each other out and a reasonably accurate estimate of the true situation will be obtained. By the time the survey has been assembled as a complete dataset some inaccuracies will be almost inevitable.

The second important temporal characteristic of geographical data is the frequency with which a given class of phenomena is measured on either a population or sample basis, in respect of the same or perhaps slightly modified variables or attributes. This issue is important because phenomena change over time and the rate of change is itself variable and perhaps unknown. Repeated data gathering in respect of a particular class of phenomena enables a time series to be built in order to monitor and examine processes of change; however, each dataset is the result of a discrete data collection event. Thus, in the case of samples and even statistical populations, different individual members of the class of phenomena are likely to be included at different points in time. This problem clearly means that like is not necessarily being compared with like.

The temporal referencing of a geographical dataset essentially involves assigning a date to the period during which data collection takes place. Thus population censuses, which are held in most countries in the first or second year of a decade (i.e. 1980 or 1981, and 1990 or 1991), are complete counts of a defined population base in that year. The importance of the census for accurately enumerating the entire population results in some countries associating the date of the census even more specifically. For example, the 1991 Population Census in the United Kingdom related to the night of 21/22 April 1991. In the earlier example of datasets derived from weather stations, recording variables and attributes which are in a constant state of flux, the time interval between data collection events is much shorter, perhaps as little as 6 hours or even less. In contrast to these examples, when some geographical datasets are generated the temporal component may seem to be relatively unimportant. One-off surveys of residents in a commuter village or of sand dunes behind an embayment appear to fall into this category, nevertheless both surveys will have taken a certain length of time and will have occurred in a particular year, month or day. If any re-examination of the same phenomena in these localities were to be undertaken on a subsequent occasion then the temporal aspect of the initial surveys might need to be reviewed.

The temporal dimension received relatively scant attention during the early developmental years of geographical information systems in comparison with the spatial component. Unlike spatial referencing, the temporal reference associated with a geographical dataset, let alone with the data records for individual phenomena, is unlikely to be held digitally as an integral part of the data. The temporal component is more often held in the documentation for a dataset or simply mentioned in its title. Where a digital temporal code has been used, this typically takes the form of a date or time such as 07/11/94 or 19:30:43. The precise format of such date and time codes may vary between software and reflect national conventions. The opportunities for manipulating such information, for example to calculate the length of time between two separate measurements of a phenomenon for the same variable, are becoming more widely available. As longer time series of geographical datasets build up and as automated capture of digital data spreads, allowing the incorporation of a date and time code from the computer's clock, then it is likely that the situation will change.

Chapter 2

Generic Problems of Data

2.1 POPULATIONS AND SAMPLES

Some reference has already been made to populations and samples in the previous chapter. Although this is not a statistics text, and several exist which describe the difference more fully than can be justified here (Shaw and Wheeler, 1985: 35–36; Walford, 1995: 24–36; Flowerdew and Martin, 1997: 93–99; Robinson, 1998: 28–31), some further explanation of the distinction between the two types of data collection is necessary. Essentially the definitional difference between a population and a sample is quite simple: a population is all occurrences of a given class of phenomena, whereas a sample is a subset of them. In principle, if a population of phenomena is available and can be measured without too much difficulty or cost, then it is sensible to do so. Unfortunately there are a number of situations in which members of a population are inaccessible or too numerous so that collection of the desired variables and attributes cannot take place. An added complication in the case of geographical data is that the spatial component may be complete (e.g. the digital boundaries for all administrative polygons), but the attribute component may be only partial (e.g. a sample survey of lone parent households).

The difficulties inhibiting collection of data from a population can be considered from both theoretical and practical perspectives. Some populations exist in theory but are inaccessible or difficult to define in practice. A number of classes of geographically distributed phenomena, such as households in a country or sand dunes on a coastline, are in a state of constant change. In other words the population, although theoretically capable of definition and enumeration at a single instant in time, has for practical purposes a certain fluidity in its membership. Households form and dissolve, sand dunes develop, migrate and disappear. In other cases the population may relate to a certain set of historical phenomena, which existed in its entirety in the past, but over time some occurrences have become lost, obscured or destroyed. In this situation the members of the population which the researcher can investigate are those which have survived the ravages of time through to the present day and it may be questioned whether this subset has been privileged in some indeterminate way. It is the available population as opposed to the theoretical population.

Even when all members of a population are potentially available and accessible for investigation, other types of practical problem can arise which inhibit collection of data from the entire set. The two most common problems relate to the time and cost it would take to collect the desired data from a complete population. The question of the time taken to collect data is usually dictated by considerations of expediency and by the need to ensure that changes to members of the population over the data collection period are minimized (see Chapter 1). Whatever means are used, data collection costs money and in most circumstances there is a finite amount available for this aspect of a project, which has to be balanced against other elements, such as data processing and analysis.

In view of these issues, data are usually collected for a sample selected from the parent population. Two questions then arise: how to select a sample; and how many observations to include in it. There are various methods by which a sample can be drawn from its parent population, however these divide into the two broad categories of probability and non-probability sampling methods (Walford, 1995). From the purist point of view, non-probability methods are usually regarded as introducing an unacceptable level of bias and subjectivity into the selection and thus tend to be denigrated for scientific investigations. Probability sampling, on the other hand, being founded on the notion of random selection attempts to encapsulate in a representative way the characteristics of a population. These include any randomly distributed aberrant occurrences as well as the underlying pattern. The simple random sample represents probability sampling in its purest form and gives all members of a population an equal chance or probability of being included. Thus aberrant occurrences are as likely to be included as those which possess the more usual characteristics of the population. The principle underlying the argument in favour of this approach is that bias is avoided and the characteristics of the typical occurrences will be sufficient so as to overwhelm the untypical ones. In order to determine whether this is the case, the array of techniques known as inferential statistics are available to test the probability of having obtained any particular result from the sample data by chance or sampling error.

Sampling of populations of geographical phenomena is undoubtedly important, particularly when scientific research investigations are undertaken. However in many applications of GIS to real-world situations, such sample data may be considered inadequate. Part of the problem relates to the geographical scale at which the sample survey is valid. Some surveys can legitimately be analysed and interpreted at national, regional or even county scales, but are unreliable for smaller spatial entities, such as UK counties, parliamentary constituencies or electoral wards. A more fundamental reservation about sampled geographical data is that many definitions of geography as an academic discipline refer to its concern with the incidence of and interrelationships between features occurring on or near the Earth's surface. While it is quite feasible for academic geographers to be concerned with a sample of a particular class of phenomena, it would be very unhelpful for a functional GIS to include only a selection of entities. For example, little would be gained from having a geographical database of a transport network which only included a random sample of roads, railways and other features.

There are some situations in which even though the data are collected from the entire population only a sample of observations are processed or analysed. The modern Population Census in the United Kingdom has hitherto provided an example. From a data processing point of view, the questions on the census schedule up to the 1991 census have fallen into two groups, those which are digitally encoded for all returned forms and the subset of questions, including those relating to a person's education, occupation, employment and mode of travel to work, which are captured for a 10% sample of forms. This approach might be adopted when the cost of discarding this information is less than that of devising and distributing separate forms to a sample of households.

Similar situations may arise in the case of collecting digital map data to represent the location of geographical features for use in computer mapping or GIS software. Although it makes little sense to sample the features themselves, it might be appropriate to do so with respect to the co-ordinates used to represent a given occurrence of a feature. When digitizing a linear or area feature each change of direction of a line could be recorded by means of a new pair of co-ordinates at each vertex (Figure 2.1a). If the lines forming the feature are very sinuous, then a large number of co-ordinate pairs will be required. This can produce very large digital map files, which are wasteful of computer storage space and processing time. One solution to the problem is to sample the co-ordinates at either the digitizing or map display stages. In the former case, the method for sampling depends on whether a manual or automatic digitizer is used. Manual sampling involves the operator deciding, usually subjectively, on which points to capture (Figure 2.1b), whereas automatic digitizers have the facility to capture points at predefined time or distance intervals (stream digitizing), for example every 5 s or every 5 mm traversed along a line (Figure 2.1c). Whichever method is used, the process of sampling defines the maximum detail that can be displayed at the visualization stage of GIS processing.

In the case of sampling at the display stage, the software eliminates those co-ordinate pairs which would detract from the clarity of the map. Again this weeding may be achieved in two ways, either by assigning a code to each vertex which denotes its importance when they are digitized (Figure 2.2b), or by the display software identifying the overall trend of a line and then automatically omitting to print or plot those vertices and their connecting arcs which lie outside a chosen distance of this trend (Figure 2.2c). The effect of sampling points in digital map files at either the data capture or display stages is to simplify the lines, to speed up the processing of digital files and to reduce data redundancy. It should, however, be noted that the generalization achieved by this operation may distort the information contained on the map, which can have serious implications for spatial searching, overlay and similar analytical operations.

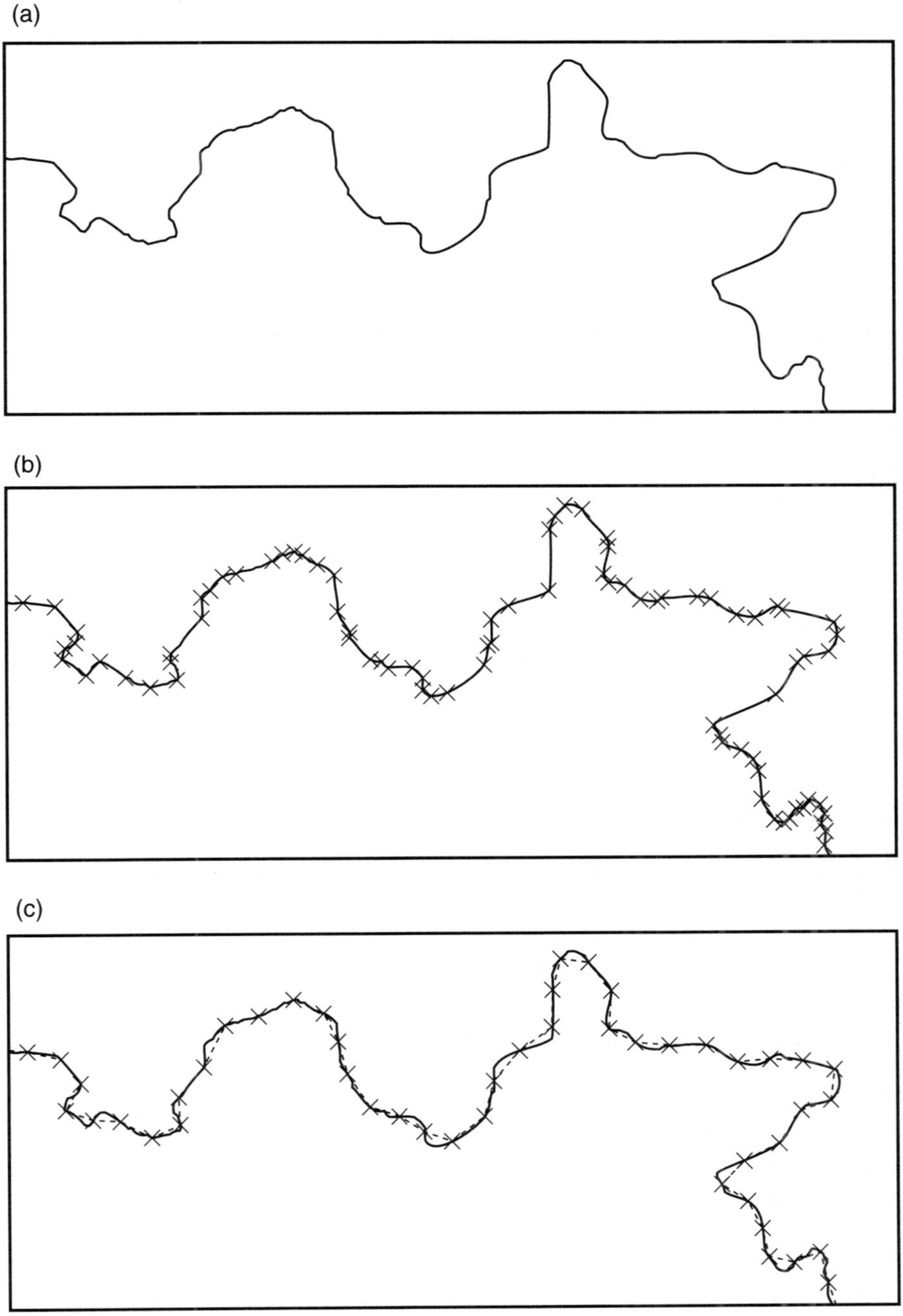

Figure 2.1 Sampling digital co-ordinates on capture. (a) Capture of co-ordinates pairs at each vertex, (b) operator manual sampling, (c) automated regular interval sampling.

(a)

(b)

(c)

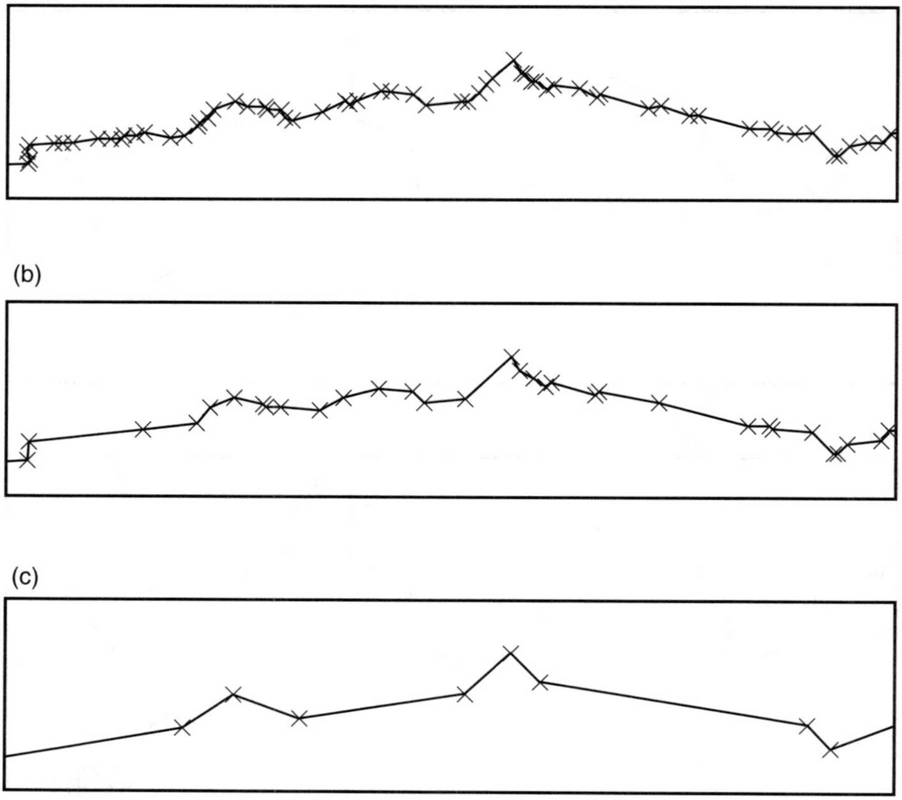

Figure 2.2 Sampling digital co-ordinates for display. (a) All co-ordinates used for display, (b) co-ordinates displayed according to coded importance value, (c) co-ordinates displayed according to trend line.

2.2 DATA ACCURACY AND PRECISION

The process of sampling data, whether in respect of occurrences of geographical phenomena or the number of co-ordinate pairs used to depict a line, inevitably introduces the possibility of sampling error which may deflect the analytical results obtained from their 'true' population values or distort the visual appearance and geometry of geographical features. Sampling error is a problem that is taken into account in scientific investigations by use of inferential statistical tests. This section is concerned with other ways in which the accuracy and precision of geographical data can be examined. There are two main types of errors found in geographical datasets, either positional error, in which some feature has been incorrectly placed or located, or thematic error, in which case there is a mistake in one or more of the attributes or variables associated with a particular feature. These are referred to as 'locational' and 'descriptive' errors by Bedard (1987) and as 'positional' and 'attribute' errors by Chrisman (1987). Both types can contribute to inaccuracy in geographical databases.

The concept of data accuracy implies that there is some absolute true value for any variable or attribute in respect of a particular phenomenon. From a theoretical perspective this is an entirely reasonable proposition, surely few people would query the accuracy of the values obtained for a particular variable when it was measured in a rigorous and careful fashion. However, it is in the nature of variables to vary. Consequently, while the values for a given collection of phenomena may be true at a certain point in time, on another occasion some or all of them may have altered. Thus the recorded values are no longer accurate. Absolute data accuracy is therefore an ideal which can rarely be achieved in practice and it is perhaps better, therefore, to think of accuracy as a relative concept.

There is a similar problem with the notion of precision, which applies to both continuous and discrete measurements. Continuous variables are usually quantified in terms of either a measurement unit, such as those relating to distance (m), temperature (°C), volume (m^3), area (km^2) and currency (£), or a ratio, such as parts per million, persons per room and live-births per 1000 population. These quantities could be recorded to an infinite number of decimal places, however for practical purposes such a high degree of precision is unnecessary. A similar problem, although on a less extreme scale, may arise with categorical or discrete attributes where the classes are part of a hierarchical classification. Industrial classifications often operate on this basis, for example at the top of the UK's classification are industrial sectors, each of which subdivide into industrial groups, which in turn split into industrial classes. The precision with which any person is categorized with respect to their industry of work depends on the level in the hierarchy.

Statistical texts often refer to the categories of coded attributes as being defined unambiguously with each occurrence of a phenomenon capable of allocation to a single class (Williams, 1985; Walford, 1995). However this is not always as simple as it might seem. When defining classes a decision has to be made about the level of precision that is required. Part of the problem is to devise sufficient classes so that satisfactory discrimination between the observations is achieved, but not so many that too few occurrences are assigned to each. It is also necessary to define classes that are meaningful. For example, housing tenure frequently occurs as an attribute in censuses and social surveys, but which tenure categories are needed? Owner occupation seems an obvious candidate, but owner occupier households can be divided into those owning outright and those buying their housing with a mortgage. Furthermore, different types of mortgage are available, so it may also be necessary to distinguish between capital repayment and endowment mortgages.

Absolute accuracy and infinite precision may be theoretical goals in the collection of geographical and other forms of data, but they are in practice unattainable targets. This does not mean there is an open season on inaccuracy and imprecision, but simply that there will always remain a chance that the data will contain errors despite the surveyor's best endeavour. In these circumstances the best advice given to the collector and processor of data is to avoid any obvious and unnecessary sources of error creeping into a dataset and where feasible attempt to quantify the magnitude of the error. One of the important points to emerge from Veregin's (1989) discussion of 'modelling error in GIS operations' (4) is his emphasis on the fact that inaccuracy in geographical datasets may not simply be associated with the collection of data, but may also be propagated by the operations carried out on the data. Returning to the line digitization example in Section 2.1, it is relatively easy to see how some of the sinuosity of a digitally encoded line may derive from operator error, where fatigue, poor manual dexterity or simply carelessness has resulted in a relatively straight line becoming more convoluted than necessary or *vice versa*. But equally important is the less obvious error associated with the processes of digitally encoding and with the operations carried out on spatial data. This latter type of inaccuracy is not the result of mistakes in data entry or processing, but arises from the nature of the digital data and the operations being carried out.

Datasets may appear relatively error-free when examined individually in terms of their internal validity (see Section 2.3), but they can produce or demonstrate errors when overlaid on each other or compared with the cartographic input. Lines on two separate data layers, which should be coincident in their geographical location, may have been digitized separately and possess some deviation. Suppose for example a stream is used as the administrative boundary for a local government area. On the printed source map these might be shown as identical with each other (Figure 2.3a), whereas when the digital map layers are overlaid, differences between the course of the two lines could be revealed (Figure 2.3b). It would be quite reasonable to digitize administrative and water features into separate layers which then might be overlaid, only to result in spurious polygons (Figure 2.3c) as well as mismatched lines. In both these cases there is a presumption that the cartography from which the digitized map layers are produced is itself accurate, which may not of course be a valid assumption.

The map overlay operation can also propagate other types of error, for example through differences in the conceptual rigour of the mapped features. All maps are an abstraction or analogue model of reality in the sense that the cartographer attempts to depict a selection of real-world features in a way that is clear and easy to interpret. However, some features shown on maps are more abstract than others. This typically occurs with polygonal features which need to be defined with more precision for cartographic display than their conceptual nature would warrant. Some features, such as soil and climatic zones, have fuzzy or imprecise boundaries which are forced into a rigid representation for cartographic purposes. If a digital map layer of such polygons is overlaid on one containing more precisely defined features, such as building plots, a different type of error is produced. Some sources of error are more easily modelled or quantified than others. One early attempt to summarize the different approaches (Veregin, 1989) identified approaches for map overlay, cartographic and thematic error, the latter separately distinguishing between errors associated with numerical and categorical variables and attributes.

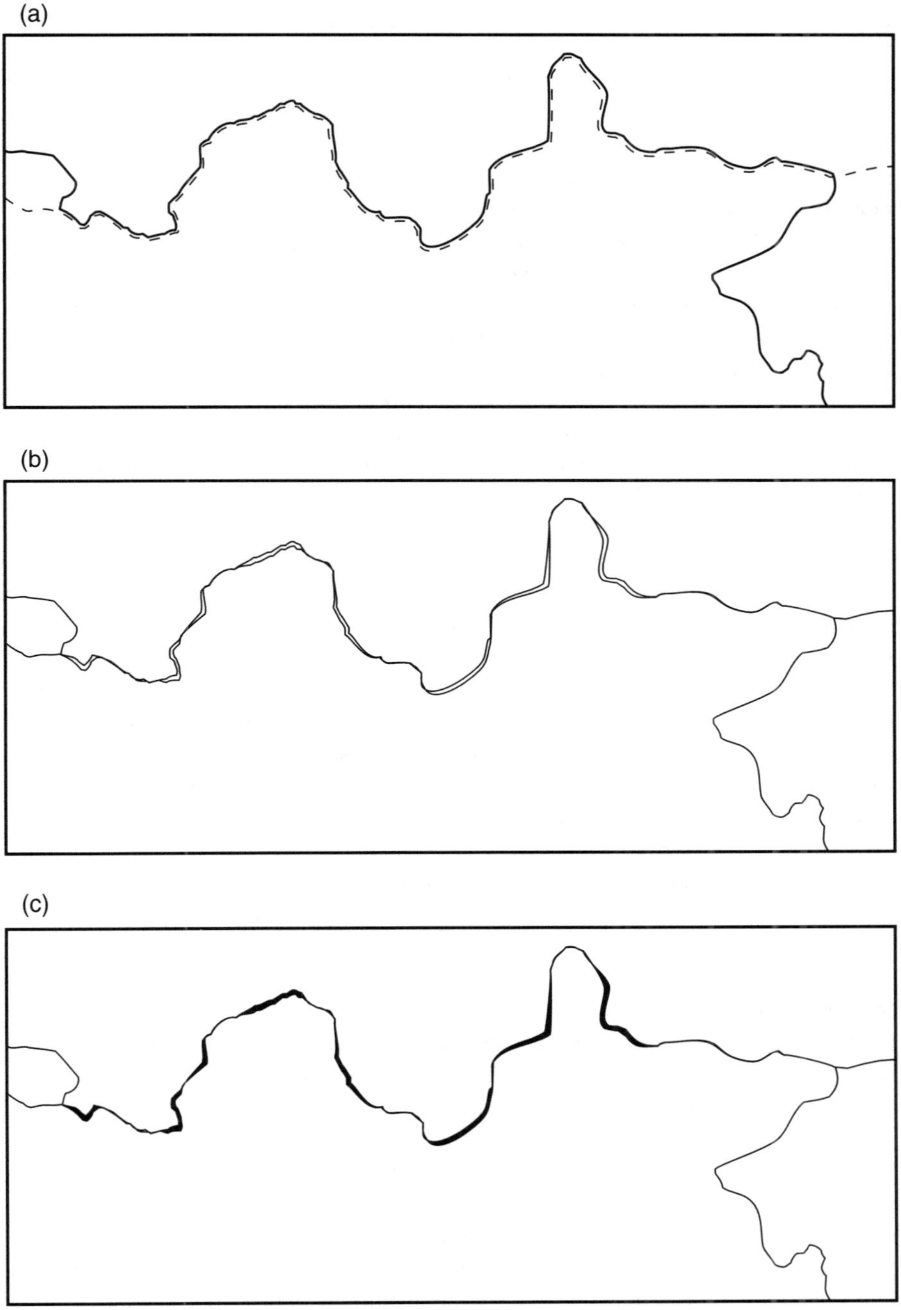

Figure 2.3 Sources of digitization error. (a) Coincident administrative boundary and stream, (b) differential digitizing of administrative boundary and stream, (c) spurious polygons arising from overlay of differential digitizing.

2.3 INTERNAL AND EXTERNAL VALIDITY OF DATA

The internal and external validity of data relates to the question of consistency both within individual cases or records in a dataset and between one dataset and another. In the former case, the essence of the concern is that two similar or supposedly identical data items should give the same information. Some questionnaire surveys, particularly those carried out by means of a face-to-face interview, will seek to elicit the same information in two different ways on the questionnaire schedule. For example, respondents might be asked, at an early stage in the interview, to indicate whether they had visited a selection of leisure facilities during the previous week. Later on they might be asked to provide a diary of their activities during this period. How any inconsistencies in the responses to these questions are treated is open to some debate. It could be argued that the diary responses were more thought out and therefore likely to be the valid response, but this may not be the case. Any difference between the answers to duplicate questions also introduces some uncertainty over the validity of other data obtained from a particular respondent on which an internal validity check has not been made.

Another form of inconsistency within a dataset can occur when different parts of the data are collected at different times or by different people. One example is provided by the topographic maps. When such maps are published for the first time, they are likely to have been generated from field or aerial photographic surveying which was carried out over several years, thus the survey data may be inconsistent for separate map sheets, which represents a particular example of the temporal data problems already discussed. However, perhaps a more serious problem arises when topographic maps are reissued, but only the part of a map sheet in which major changes are believed to have taken place is resurveyed. In this case there is inconsistency within the same dataset or map sheet. The published maps may indicate when and for which sections of the map the resurvey was undertaken, however for the inexperienced user the need to search out this piece of information may not be apparent.

Since data from the first Earth observation satellites became available to researchers (see Chapter 4) in the 1960s, considerable academic and applied research effort has been expended in using such data to determine vegetation and land cover in both urban and rural environments (Nagel and Robinson, 1990). Collecting data on these topics at a regional, let alone national or international scale, by means of field observation is an extremely costly and time-consuming operation, which may be prone to errors of classification and measurement. One significant benefit of remote sensing is to spread the cost and at the same time increase the periodicity and currency of such information. The procedure for processing the remotely sensed data will involve a certain amount of 'ground truthing' in order to corroborate the land use and vegetation information gained from the satellites with field survey results. Even if this is carried out satisfactorily, it is possible that inconsistencies between the two types of data will remain unresolved.

Chapter 3

Accessibility and Availability of Geographical Data

3.1 INTRODUCTION

One issue is becoming increasingly important as the use and complexity of information technology, and of geographical information systems in particular, develop. The issue has many aspects and relates to the accessibility of data, particularly from secondary sources. It is nearly 10 years since Cassettari suggested that we are moving into a situation of 'being data rich but information poor' (Cassettari, 1993: 7). One of the effects of the growing dependence on information and communications technology has been to heighten people's expectations and requirements with regard to obtaining geographical and other types of data for their use. This expansion in demand is being met by an enhanced supply, which are together highlighting a set of 'access issues'. Questions of access can usefully be grouped under four main headings: legal matters including ownership, copyright and confidentiality of data; technical issues relating to the means of data access; the location and sources of data; and practical problems concerning matters such as the cost, characteristics and location of data sources. The purpose of this chapter is to examine these overarching issues relating to the accessibility of geographical data in order to identify common themes which transcend the particularities associated with types of data or countries. Before examining these general access points, it is worth reflecting on the basic question of why users of geographical information systems might need to access data.

The simple answer to this question is that GIS are information systems which would have little practical value and would have developed less rapidly and extensively, if they were not intended to be used with data relating to practical 'real-world' geographical problems. GIS are therefore predominantly systems for capturing, manipulating, storing, analysing and displaying data which are referenced to the Earth's surface (HMSO, 1987). Furthermore, many of the problems to which GIS may be applied entail the integration of data which are differentiated in terms of their source and/or substantive content. Such problems may be of various types and could be placed into different categories, for example locational decision-making, resource assessment and facility management. A number of standard texts already demonstrate the now extensive range of applications in which GIS are employed in both the developed and developing worlds (Ball and Babbage, 1989; Maguire *et al*, 1993b; Martin, 1995). As the desire to expand the range of applications continues, issues of data accessibility are emerging as crucial for the further development of the technology.

It is a consequence of the information technology 'revolution' that current and potential users want and expect access to greater quantities of up-to-date information. A feature of living through this rapidly changing information era is a general perception that the rates at which data are generated and consumed have accelerated. There is, however, nothing new about an increasing demand for information, it has occurred since societies first started to organize themselves and the governing individuals or institutions found they needed information about the condition of their populations and their environment. It is apparent that not only governments want information, but also trading and commercial organizations seek knowledge of different environments as potential sources of raw materials and about markets for their goods and services. If history tells us that the need for information has existed for a long time, what is new in the current situation? The threefold answer is increased diversity in the range of information sources, an improved facility for integrating and visualizing different types of data and, pre-eminently, the greater ease with which data can be moved between one individual or organization and another.

The historical development of the demand for information reveals one important lesson for those caught in the contemporary maelström of the information revolution. The demand for and the supply of data varies according to the kinds of people and organization on both sides of the 'information equation'. As a result, different people can access the same data in a variety of ways, depending on a number of factors including the facilities available, the amount of detail required and the purpose for which the information will be used. In recent years the growing complexity of these issues and the wealth of data available has led to the commodification and packaging of data to suit particular purposes. Commodification is the process whereby datasets, or products derived from them, have become saleable commodities, while the packaging of data relates to the process of assembling subsets of observations and/or variables from one or more datasets. These processes are often related and have resulted in both public and private sector organizations assembling products to suit different data users' needs. From the perspective of simplifying access to complex datasets, this approach has much to commend it, however there are some drawbacks which may not be immediately apparent to users who are unfamiliar with the data in the first place.

The commodification and packaging of data have a number of consequences for the access issues considered in the remaining sections of this chapter. Although many users still want access to unprocessed data, the scene can be set for looking at access issues by considering a series of rhetorical, but nonetheless realistic, questions. For example, what happens to the primary data collector's copyright on raw data when these are manipulated by statistical or other means into composite indicators or even simple ratios or percentages? What are the implications on copyright of converting data which are stored on one medium (e.g. paper or videotape) into a digitally encoded form for storage on a computer, for example statistics taken from a published volume or geographical features from a printed map? When the information encapsulated in raw data is simplified into a packaged product, for example, categories are collapsed into a smaller number of groups or the digital boundaries of geographical features are simplified (generalized). In these circumstances, how can disputes between users of the original and the packaged information be resolved? Different interpretations may be made from the two sets of information which purport to record the same phenomena. Organizations involved with facilitating access to raw, unprocessed data and with assembling package products are acting as brokers between the primary collectors and the secondary users of data. How should this added-value be rated and what royalties should be remunerated?

Access to data is clearly a multifaceted question, which has been distilled into four broad groups of issues. The relative importance of these issues will vary from one dataset to another and from one user-environment to another. However, the diversity of data sources currently or, more importantly, potentially available for use in geographical information systems means that access questions cannot be ignored. It is significant that organizations have been established to represent the users and providers of geographic data and GIS. In the UK the Association for Geographic Information (AGI) in its current form and the government's Interdepartmental Group on Geographic Information (IGGI) were founded in 1997 and 1993 respectively, whereas the research orientated National Center for Geographic Information and Analysis (NCGIA) in the USA was set up in 1988. Many of the access issues considered here have been or are currently on the agendas of these organizations, which have for example been instrumental in devising standards for digital geographic data.

3.2 LEGAL ISSUES

One of the most fundamental, but perhaps also most perplexing, limitations on access to geographical and other types of data from secondary sources relates to legal issues and people's willingness to make information available. Dale and Marsh highlighted a key arena of conflict for modern societies, namely 'the right to privacy and the right to information' (Marsh, 1993: 115). They further suggest that the organization of societies presupposes 'the availability of accurate information on the population and its characteristics' (Marsh, 1993: 115), to which might be added information about the physical environment inhabited by that population. A series of problematic areas are relevant to this discussion, including data confidentiality, data copyright and ownership of intellectual property rights. The data collector or owner and the actual or potential data user have contrasting perspectives on these legal issues. The arguments on both sides are complex and positions change as various forms of pressure are brought to bear. A further complication is that differential access arrangements may depend on whether data have been collected with private or public finance.

Some data may be of no use or consequence to anyone other than the primary data collector either now or in the future. However, in practice it is difficult to envisage every possible use that may be made of a particular set of data and therefore the question remains whether each dataset ought to be preserved, in much the same way that a national library holds a copy of published items. This raises the question of the point at which data can be regarded as published. The British Library does not hold copies of confidential reports produced for companies and government papers are usually only released after 30 years and in some cases longer. In the information era the first attempt in the UK to legislate for issues of data confidentiality came in the form of the 1984 *Data Protection Act* (Data Protection Registrar, 1992), which has been updated by new legislation that came into force in 2001. Its principal focus was on protecting individuals from inaccuracy in information held about them on electronic media and established a person's right of access to view data held about them (Table 3.1). The UK legislation has generally followed the precepts of the Council of Europe Convention concerning data confidentiality which states that 'data should be: obtained fairly and lawfully; used only for specific legitimate purposes; adequate and relevant for specified purposes; accurate and up to date; kept only as long as needed for specified purposes. Any person should be able to: establish the existence of a file containing personal data; obtain details of any data held on themselves; have details rectified if they are inaccurate; obtain remedy if the data users do not

Table 3.1 Principles of 1984 *Data Protection Act* in the UK

1. The information to be contained in personal data shall be obtained, and personal data shall be processed, fairly and lawfully.
2. Personal data shall be held for one or more specified lawful purposes.
3. Personal data held for any purpose or purposes shall not be used or disclosed in any manner incompatible with that purposes or purposes.
4. Personal data held for any purpose or purposes shall be adequate, relevant and not excessive in relation to that purpose or purposes.
5. Personal data shall be accurate and, where necessary, kept up to date.
6. Personal data held for any purpose or purposes shall not be kept for longer than is necessary for that purpose or those purposes.
7. An individual shall be entitled at reasonable intervals and without undue delay or expense
    - to be informed by any data user whether he[or she] holds personal data of which that individual is the subject;
    - to access any data held by a data user; and
    - to have such data corrected or erased, where appropriate.
8. Appropriate security measures shall be taken against unauthorized access to, or alteration, disclosure or destruction of, personal data and against accidental loss or destruction of personal data.

Source: Data Protection Registrar, 1992.

provide details of data held or fails to rectify inaccuracies' (Peers and Buckley, 1988).

The question of confidentiality becomes particularly acute when there is a statutory requirement for the general public to provide information to government. This can be highlighted by the case of the Census of Population, which is probably the most well-known example because it relates to every person living in a country at a particular time where there is a statutory requirement for compliance with providing the requisite information. Alonson and Star point out that '. . . growing public anxieties about invasion of privacy [led to] cancellation of West German censuses in the early 1980s', since there was 'a threatened statistics strike by about one fourth of the population' (Alonson and Star, 1987). The UK census authorities have striven to allay such fears and latent opposition by providing a clear confidentiality statement.

It is important to distinguish between the theoretical and perceived risk of disclosure of confidential information. The measures taken by the UK census offices to distort or 'Barnardize' the statistical output by the semi-random addition of -1, 0 or $+1$ to cells in the tables, whilst maintaining their internal consistency and the use of thresholds to disallow the release of figures for areas with small population or household totals, means that the data users' access is not precisely what was provided by members of the population. These adjustments reduce the theoretical risk of disclosure to an infinitesimally small amount. However, of greater importance than this theoretical risk is the public's perception of whether the information that they provide will get into the hands of people outside of the census offices. The UK government has a strong commitment to preserving the confidentiality of population census information. Census confidentiality is governed by the principles contained in Table 3.2. Bulmer (1979) claimed that 'confidentiality involves the conditions of use and disclosure of information about individuals or businesses once it is collected', hence, although we have seen greater possibilities of data access in the last two decades, this has not been without the price of needing to conform to and perhaps sign agreements relating to conditions of use.

The question of copyright over data and the intellectual property rights pertaining to the information contained in data seems an evolving and continuing debate in respect of electronic media. It is perhaps the easiest thing in the information technology world to make a copy of data on a computer; when obtaining data for analysis in GIS or other software we frequently intend to make copies for our own use, indeed the process of analysis itself is likely to multiply the number of copies of a particular data file. Furthermore, users are generally encouraged to make backup copies of data held on their systems to avoid the problems that arise from accidental or malicious loss or damage. It might seem reasonable to assume that legitimately obtained data that are transformed by a user are no longer subject to the original copyright and intellectual property rights, but, unfortunately, this is not necessarily the case. In legal terms 'copyright is a *property right* that gives the owner exclusive right to perform certain acts (termed *restrictive acts*)' (Wadsworth and Treweek, 1999: 53–54, original emphasis). As the ability to obtain geographical information becomes easier, for example to view and print out street maps over the internet, the boundaries of where the original owner's copyright ceases remains uncertain. Copyright owners are often

Table 3.2 Principles of confidentiality in the UK Census of Population

1.	While in most cases one form will cover all members of the household and it will be collected by the enumerator. People may, if they wish, be able to give personal information in a way that will not reveal it to others in their household or establishment or to the local enumerator.
2.	All members of the Census organization will be given strict instructions about confidentiality, and will be liable to prosecution for any breaches of the law. Access to personal census information within the Census Offices will be strictly limited to those staff needing to see it to carry out their official duties.
3.	The physical security of Census documents containing personal information held in the Census Offices will be strictly observed.
4.	The computer system handling Census data will have safeguards that prevent unauthorised access to the information held. In particular, *names and addresses* will not be stored in any computer.
5.	Information from the Census will be used solely for statistical purposes. As in the past, the Census Offices will not pass information about identified persons or households from completed Census forms to other government departments or to anyone else outside the Census organization. Moreover, in releasing statistics from the Census, all possible steps will be taken to prevent the inadvertent disclosure of information about identifiable individuals.
6.	The returns themselves and the information they contain will not be made available to the public until 100 years have elapsed, when the confidentiality of the information is held to have expired. This period of closure was prescribed in an Instrument of the Lord Chancellor in 1966 under the Public Records Act 1958.

Source: HMSO, 1991.

willing to allow the use of data for educational and perhaps research purposes without payment of a royalty, but when commercial gain is on the horizon, for example when academics undertake contract research or publish material in books, this willingness may evaporate. The safest assumption to make is that data are subject to copyright unless there is some clear indication that this is not the case.

3.3 MEANS OF ACCESS

Access to geographical data may be achieved in various ways: this section presents an overview of how data may be accessed. It has already been established that the term 'geographical data' is broadly defined and includes not only data which have already been captured in a digital format (i.e. held in a machine-readable form on electronic storage media), but also those which have the potential to be processed and stored in this way. The breadth of this definition is important since many geographical data are held on a printed or photographic media and are rendered into a digital form for processing in a GIS or other data handling system. A related issue concerns whether the user of the data is able, or needs, to obtain the original version of a particular data collection, for example historical records or survey documents, or whether a digital copy will suffice. We can thus define access to geographical datasets as the ability or opportunity for an individual or organization to make use of data, subject to any legal conditions of the type discussed in Section 3.1. The problems and issues connected with access refer to the technical matters and the procedures whereby data may be obtained.

The overall amount of geographical data held in a digital as opposed to a non-digital form is increasing as part of the general spread of information technology and as the development of automated data capture devices and software makes the use of such equipment a feasible proposition. Rhind (1988) suggested that there were four factors affecting the spread and development of information technology: the availability of hardware, software, liveware (people) and data. He argued that the relative importance of these factors was changing as improvements in computer technology and training took place. He envisaged that the first three factors would largely have been conquered by the end of the 1980s. In the 1990s, proliferation in the amount of data being collected would constitute the fourth wave of problems to be overcome as people were initially overwhelmed by the exponential growth in the quantity of data. Figure 3.1 represents the growth in geographical data since the 1960s when such information started to be processed on computer systems, and distinguishes between two types of geographical data: first, those existing in a non-machine readable form that have to be entered for computer processing, and, secondly, those already held on or captured onto electronic media in a digital form.

In the early years, before computers became widely used for geographical data processing, paper-based information was more important apart from in the area of satellite remote sensing. Many person-years have been expended on converting statistical and cartographic information stored on paper and other similar media into a digital form before any analysis can be undertaken. Within the last 10 years the direct capture of data has become more significant. As the information technology era rolls inexorably forward, then it may be anticipated that more and more geographically referenced data will be captured directly onto a computer system.

The distinction between non-digital and digital data provides a convenient starting point for examining means of access, since it suggests an initial division between computer-based access routes and those involving the preliminary step of obtaining data from archives, libraries, commercial companies or other repositories of non-digital information. Certain of these organizations may also hold some or even all of their geographical data on machine-readable media, and thus some general access issues transcend the digital/non-digital divide. Non-digital data may sometimes only be accessed in their original form, particularly in the case of historical documents and photographs. Consequently access to this type of information may be difficult and involve painstaking research to find out what is still in existence, let alone whether it may be accessed.

3.3.1 Means of Access to Non-digital Data

The main question relating to the means of access to non-digital data concerns the identification of where such data are located and in what form the information is held. Part of the problem with locating such data is that they tend to be held in a

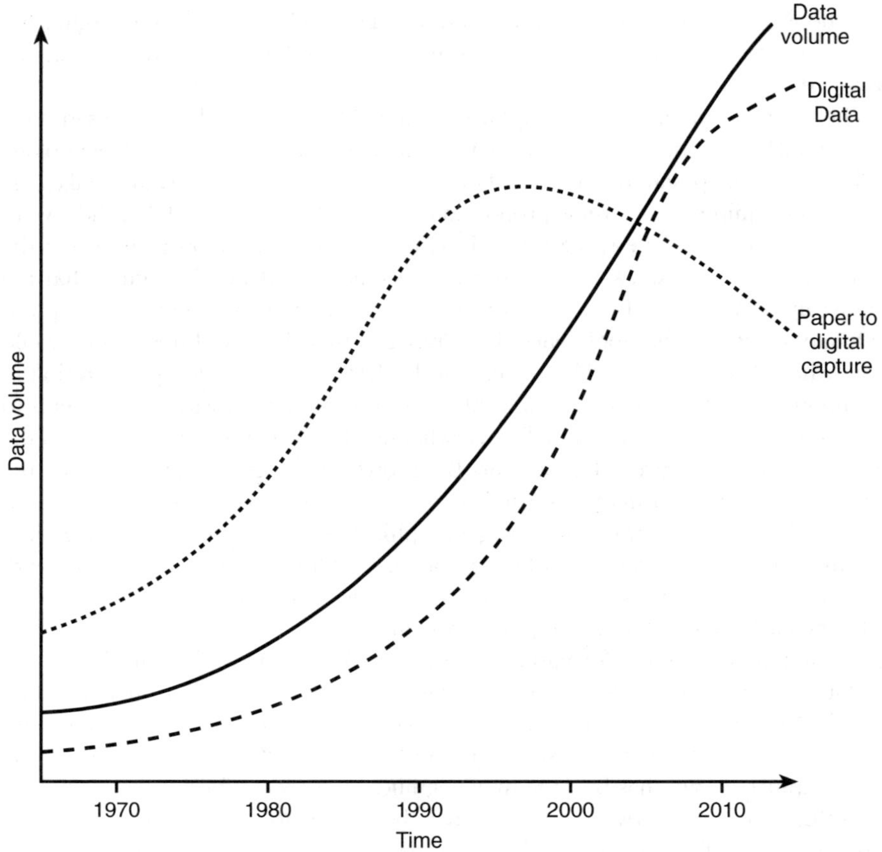

Figure 3.1 Trends in quantity of paper-based geographical data capture and digital data since 1970s.

relatively decentralized fashion. Organizations which hold non-digital geographical data range from national to local, and are either in public or private ownership. National public sector organizations in the UK include the Public Record Office, British Library and the Royal Geographical Society. Similar organizations exist at the local level in the form of county record offices and libraries. From the perspective of the geographical researcher, the searching out of new sources of information is often a significant part of the process. However, when the intended analysis is of a non-academic nature, there may be less time available to track down the obscure and more interest in using 'tried and tested' sources.

3.3.2 Means of Access to Digital Data

The key question relating to the means of access to digital data concerns the readiness with which the data can be used with the hardware and software, the geographical data handling system, available to the user. There is a close connection between the format in which data are stored and the characteristics of the user's software. There is little gain for the user if a large number of digital datasets are readily accessible, but the manner in which the data are held is incompatible with the available applications software. Conversely, the most analytically sophisticated software is of little benefit if the format of its data storage is excessively idiosyncratic and inhibits the straightforward interchange of data between different systems. During the course of the last 30 years there has been a progressive narrowing of the gap between these extreme positions. However, it is only since the early 1990s that the exchange of digital data between word-processing, spreadsheet and database software has become a commonly available facility of such software and largely transparent to the end-user. The notion of 'open GIS' is now a developing issue with respect to geographical information handling and is exemplified by internet-based mapping systems and the facility for interchanging digital geographical data between software.

Technical questions relating to the interchange of digital data concern both the hardware and software that are used for data storage and manipulation. It is not surprising that competing companies should have attempted to establish a dominant position in the marketplace during the initial phases of the information technology era. This competition has occurred in respect of mainframe hardware and latterly for personal computers. Once such dominance was established other companies sought to mimic or 'clone' systems. Similar trends can be seen in the case of software, whereby particular data storage formats, often tied in closely to hardware characteristics, gave way to more standardized data structures. As the media for storing digital data has moved through a sequence from punched cards and tape, magnetic computer tape, disks, compact disks and now network transfer, it is possible to see how a wide range of hardware and software combinations have eventually been reduced to a more restricted set of standard features. For example, when floppy disks were introduced they were available in various sizes (8″, 5.25″ and 3.5″ diameter), recording at different densities on one or both sides and in hardware specific formats. By the early 1990s floppy disk technology had largely focused on 3.5″ disk drives capable of reading and writing at various recording densities in a standard format. Such technology itself is already being superseded by compact disks for larger volumes of data. Furthermore, file interchange utilities have now consigned the separation of personal computer systems into the IBM and Apple types to the history of information technology. These changes exemplify how information technology is not only in a constant state of adaptation and evolution, but, even more importantly, it is able to adjust cumulatively thus establishing new standard configurations upon which the next round of changes can build.

One important benefit of digital data is the potential ease with which a particular dataset can be accessed, either because copies of the dataset can be distributed to users via conventional means on computer storage media, or because multiple usage of the same dataset by different users can occur simultaneously. In the former case, distribution may be as simple as making a duplicate copy of the requisite data file onto tape, floppy disk, compact disk or other storage media and sending it through the postal system. Simultaneous, multiple usage of a dataset implies that some form of remote access system is available over a local or wide area network. Remote users of a dataset may also be able to extract part or all of it and transfer the digital data over the network to their local computer system by means of file transfer protocols. These various possibilities describe systems of distributed access to data, in other words those in which the data source and the user are in physically separate locations. However there is a distinction between access routes in which, on the one hand, both the users and copies of the dataset are distributed and, on the other, those in which the users are distributed but the data are held at a centralized location (Figure 3.2). Examples of the three models of data access can still be found, but there has been a general shift over the last 30 years towards remote access systems over networks.

This brief overview of the ways in which information technology has changed provides some lessons for considering the present and future means for the interchange of digital geographic data. In many respects these data are not inherently different from those used in non-geographical information handling systems. Many of the general technical issues relating to the storage of digital data, which have already been outlined, also apply to geographical information. Thus, compact disks and networks are now frequently used as the means for distributing geographical data to users. There are, however, two important practical differences, first, relating to the large quantity of data involved in any individual dataset and, secondly, the fact that this characteristic may be exacerbated when a particular application requires two or more datasets to be combined.

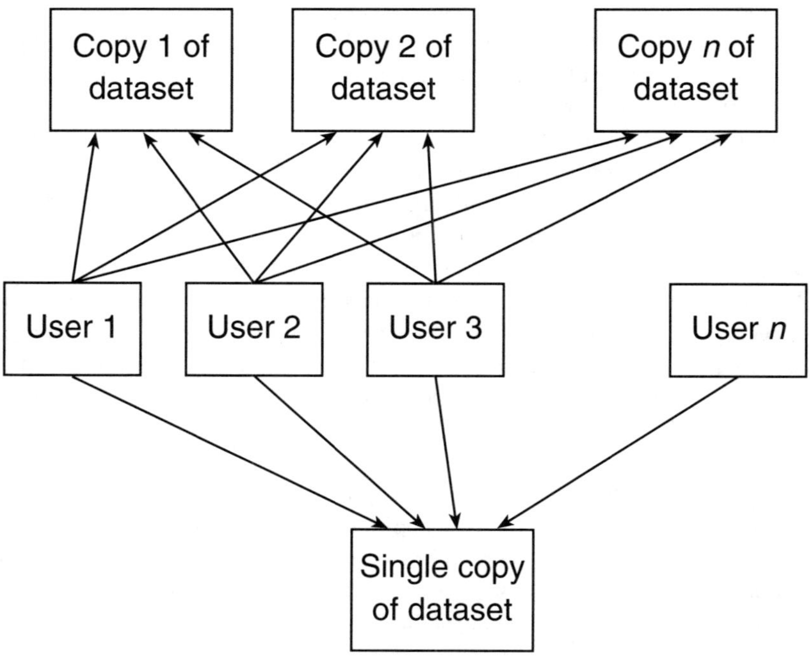

Figure 3.2 Models of remote data access systems.

The development of computer mapping and geographical information systems capable of handling large quantities of geographical data has been adequately accounted for in other texts (Burrough, 1986; Star and Estes, 1990; Berhardsen, 1992; Burrough and MacDowell, 1998). The potential for using computers to process geographical data and the development of such systems can be traced back to the early years of the information technology revolution and to some of the 'primitive' mapping software of the 1960s (e.g. Synagraphic MAPping and CAMAP). Most geographical information systems which have been developed to date have taken a cartographic approach, and thus the question of how spatial features are represented has been viewed as an important issue. The division between raster and vector forms of feature representation continues to provide a categorical division of the software, which is frequently judged on the basis of comparison with printed maps. Such evaluations may neglect the fact that traditional maps are themselves abstractions of reality and thus may be no more or less successful at adequately representing geographical features than their computer-based equivalents.

In recent years there has been a proliferation of computer mapping and GIS packages, some of which allow users to convert between vector and raster data structures, thus this problem is a less significant inhibitor of access to data than formerly. A more serious issue concerns the extent to which the format in which the digital geographical data are held is 'package-dependent'. It is now possible to see the signs of package-dependency breaking down in respect of GIS and mapping in a similar way to past changes in word-processing, spreadsheet and database software. For example, ArcView, a leading desktop and workstation GIS that operates in a Windows environment, is able to read in image files. Access to and interchange of digital geographical data has also been assisted by the development of standardized data transfer formats, such TIGER format files for US census data, the National Transfer Format (NTF) in the UK and Digital eXchange Format (DXF).

There is one crucial reason why differentiation in data formats between individual word-processing, spreadsheet and database software has largely disappeared, whereas it is likely to persist longer with GIS and mapping packages. In the latter case, users who have diverse and separate applications for GIS tools may need access to the same basic digital geographic data representing and locating physical features on the Earth's surface. For example, within a particular built-up, urban area, one user in local government with an application to determine the least damaging route for a new relief road, someone in a utility company managing infrastructure (cables and pipelines) and yet another user in a distribution firm seeking the most cost-efficient route for the delivery of goods, may all need access to a common geographical database containing information about transport infrastructure. In other words, users of mapping and GIS software often have a common need to access the same digital topographic database.

The need to access such databases not only cuts across departments within individual organizations, but also across companies operating in competition with each other. In this way basic geography and topography form a foundation for many computer mapping and GIS applications. There is no comparable, 'generally necessary' type of information which is required for word-processing, spreadsheet or database applications. The most important consequence of this situation is that, if the collectors and suppliers of the basic geographical datasets can agree to make these available in standard formats, there is less need for the producers of GIS software to encourage the interchange of data. Provided that a particular mapping or GIS package is able to accept data in a standard format (e.g. DXF or NTF) and to read attribute files in common spreadsheet or database file types, the software houses can to some extent continue to develop their products without the need to allow the interchange of digital map files.

3.4 SOURCES OF DIGITAL DATA

In the 1960s computer-based methods started to make inroads into paper-based systems as the means for storing data and for enabling such data to be retrieved and processed into information quickly and efficiently. One effect of this transition was the need to re-think the traditional ways in which data and information were preserved for both the short and medium term as well as for posterity. The paper-based archives, which constitute repositories of data stored under the formerly dominant medium, were joined by a new group of bodies variously known as data archives, data banks and data libraries. These were established either as units within individual organizations as they adjusted their record-keeping procedures or as publicly funded national centres, often located in universities and the academic sector. The essential concerns and functions of such national bodies were to preserve and disseminate statistical or other forms of data which may have been collected at public expense. Data archives within individual companies clearly serve similar purposes but operate with the aim of maintaining the private, commercially sensitive records of the company. The question of general, open access is unlikely to arise in these circumstances and the prevailing concern would be to establish appropriate restrictions on access which reflected the levels of confidentiality associated with different types of information. Some private companies have allowed academic access to statistical data, and one group in this field are those carrying out opinion poll surveys, although these tend to lack geographical detail.

One example of a national data archive, serving the academic community in the UK, was the Social Science Research Council Data Bank, which was founded at the University of Essex, Colchester, in 1967, and subsequently changed its name to Survey Archive and then Economic and Social Research Council Data Archive and now simply known as the Data Archive. The Program Library Unit (PLU) attached to the computing centre at Edinburgh University was also founded over 20 years ago, and, although initially concerned with writing and distributing software, soon extended its activities to include the development of systems allowing remote access to data. Similar archives have developed for the academic communities in other countries across Europe, North America and Australasia, including the Danish Data Archive (DDA), Norwegian Social Sciences Data Services (NSD), the Inter University Consortium for Political and Social Research (ICPSR) at the University of

Michigan and the Social Science Data Archives at the Australian National University, Canberra.

The historical development of systems for accessing digital data in the UK and elsewhere can be illustrated by reference to one particular type of data, which has acquired a key role across a range of social science disciplines as well as forming the foundation of much geodemographic analysis in the commercial and public sectors. The Population Census datasets are discussed in detail in Chapter 7, however some significant stages in the development of remote access to digital data can be gleaned from a brief account of how access to these data has evolved. The importance of census statistics is such that they have achieved a high profile, almost pioneering role in the dissemination of digital georeferenced attribute data.

The first tentative attempt to allow people outside of the former Office of Population Censuses and Surveys (OPCS) to access machine-readable statistics from the British Population Census was made during the 1960s. A limited number of counts for small areas were released on punched cards for a selection of towns and cities and in 1955 the Inter-University Census Tract Committee was formed in Oxford with the intention of defining census tracts similar to those used in the USA. As a prototype, 48 tracts were defined in Oxford with an average population of 2645 (Robson, 1969: 44). In the early 1970s, quantitative and computer-based analytical techniques had taken a firm hold of many social science disciplines and OPCS produced standard statistical summaries on computer tapes from the 1971 Census for the whole country. Extracts from this 'database' were obtained by individual academic researchers and local authorities and consequently within the various computing centres there were discrete and occasionally duplicate copies of census data on computer tape. Research in the Geography Department at Edinburgh University on computer mapping resulted in a complete set of these statistics being lodged in the University's computing centre and the PLU proceeded to negotiate with the Scottish Census authorities for permission to allow other researchers to access these data. The Survey Archive at Essex University also obtained national sets of statistics from the 1971 Population Census and the 1966 sample Census. Bespoke software was written to generate subsets of the statistics for selected geographical units for transfer to computer tape and distribution to users.

Commercial companies also became involved towards the end of the 1970s in the marketing of statistical information derived from the UK's Population Census, particularly in the packaged form of area classifications. In the early 1980s, census data dissemination took another turn when a system of remote access for the academic community was implemented, using a 'star' network of six university computing centres in conjunction with the Data Archive. The six centres provided an on-line system of access to the data using standard retrieval software, known as SASPAC. Developments in computer networking during the 1990s have enabled the data from the 1991 Population Census to be mounted in a single computing centre at Manchester University, which can be accessed by users the length and breadth of the UK.

This example illustrates a number of broad trends. First, there may be considerable diversity of sources for a particular type of data, since the organization responsible for data collection may devolve responsibility for dissemination to a large, possibly dispersed, community of users to some intermediary or third party. For example, census data for the UK can be accessed from the census authorities, from commercial companies and from public sector archives and university computing centres. Thus the source or sources of access to a particular dataset are not necessarily synonymous with the originator or collector of the data. Secondly, the organizations holding a particular type of data may only make a subset of the full dataset available or may have undertaken some form of analytical processing or packaging of the data. Although this simplifies access, it may also obscure the richness of some datasets. Thirdly, large and complex datasets often require specialized data access software to act as an interface between the raw data and the end-user. As direct access to large datasets has developed, it has been necessary to provide more user-friendly software, which can steer users through a complex series of decisions concerning areas, counts and levels of aggregation in order to reach the required part of the dataset.

The increasing demand for geographical data does not mean that the number of organizations acting as secondary sources is declining. Indeed, the reverse is the case with an expanding number of projects and bodies being established to provide data access services. The World Wide Web (WWW) has rapidly become one of the most important sources of information. After initial development for military communications purposes, it evolved into university and commercial computing networks in the 1970s and 1980s and was used for e-mail and file exchange. The WWW is now an informal collection of computer hosts which are linked together in a dynamic and, from the user's perspective, largely transparent way. Users who subscribe to a service provider are able to connect through the internet with millions of sites around the world in order to find information. An expanding number of organizations are creating host sites on the Web and using this facility to provide information about their activities, datasets and services to a worldwide audience. The essence of the Web is its openness and the fact that users simply pay to get onto the system as opposed to accessing or obtaining information. Access to certain areas (pages) at a host site can be protected by means of passwords, nevertheless the principle of the system is to allow relatively unimpeded access to information.

The legal issues of data ownership, copyright and charging (see Section 3.2) mean that, for the present at least, the World Wide Web has yet to reach its full potential as a freely accessible source of geographical and statistical data. Many organizations, which either collect primary data or act as intermediary 'brokers' between these collectors and the end-users, provide a service which adds value to a dataset. These activities usually attract some form of service charge which means that such data are not openly available over the internet on the WWW. An early example of such a service is the National On-line Manpower Information System (NOMIS) based at Durham University, which was established in 1983 to provide remote access and mapping services for the Joint Unemployment and Vacancy Office Statistics (JUVOS) which have been generated monthly since 1983 (Townsend *et al*, 1986; Townsend *et al*, 1987; Blakemore, 1990). The range of information now held as part of NOMIS has expanded to include the main national datasets containing demographic and socio-economic statistics for small areas in the UK. Access to these statistics is by means of a 'dial-up' facility which enables users to access the databases in order to display and extract statistics in mapped or other forms. Although not an entirely open access system, since users are required to register to access the system, there is nevertheless comparatively little direct contact between the service provider and the user in comparison to earlier procedures.

The development and success of systems of remote access to data depend on the nature of network communications and, perhaps most importantly, on the user interface. When procedures for accessing data through secondary sources were first established, the user interface was typically the post and/or telephone systems, often supplemented by workshops and seminars. In other words it was an interface based on personal communication and contact between the data collector or secondary source service provider and the end-user. The development of systems of remote access to data through secondary sources has necessitated the development of increasingly sophisticated computer-based user interfaces, which guide users to the information or data that they require. Developments in the NOMIS user interface typifies the progress which has occurred. When this was established users accessed datasets held in NOMIS by interactively constructing command files, which were then executed and the output, whether in mapped or tabular form, was printed at Durham University and then sent by post to the user.

3.5 PRACTICAL PROBLEMS OF ACCESS

The main practical problems for many users seeking access to geographical data from secondary sources relate to finding out what data are available, whether these data satisfy the users' requirements, and whether they are reliable. In general terms the answers to these questions are provided by what is referred to as metadata, or information about data and datasets. The assembly of such metadata is not only for the benefit of end-users but also for the personnel working in organizations which provide a data service, the data suppliers. It is therefore important to distinguish between metadata which are necessary for users to access data and those which are generated or required by organizations so that they can supply information and data to third parties. The extent of metadata needed by end-users depends on two things: first, the way in which the data are accessed, for example over a network or copied onto tapes, disks or other media, and, secondly, whether users require a discrete piece of information, the statistical equivalent of a 'sound bite', or raw data as an extract from or a complete copy of a dataset. For someone whose end purpose in accessing a dataset is simply to find out a particular statistic, for example the area of coniferous woodland in the UK's National Parks, there is little point in confusing the issue by giving details of how that piece of information is stored in the database. In contrast, for users who intend to analyse the data themselves with their own software it is vital to know how and where data are stored, and what forms of coding may have been used.

Metadata are themselves broad in scope and can be divided into three basic categories. The first relates to descriptive information about the substantive or thematic content and geographical characteristics of datasets. The second specifies the location of sources of datasets both in broad terms, such as which organizations hold a particular dataset, and more narrowly in terms of where a single data file or record is to be found. The third category is technical information, for example, details about exactly how individual data records and items are structured in digital, machine-readable form and the coding of variables. This type of metadata is required in order to make raw digital data usable by means of software. These categories can be related to the three main questions which a user of geographical data from secondary sources needs to be able to answer:

- What data are available?
- Where are they available?
- How are they stored?

The remainder of this section will examine the three types of metadata in relation to these questions.

3.5.1 Thematic and Geographical Metadata

In many cases the data that would match precisely with the user's requirements do not already exist and three options are presented to resolve this problem: abandon the quest; undertake primary collection of the requisite data; or search for data that constitute an acceptable compromise. The first seems defeatist, the second potentially expensive and time-consuming, or perhaps even impossible, and the third feasible, but not without its problems. The initial concern of potential data users when searching for a suitable secondary source is to identify datasets which satisfy their substantive or thematic requirements. The search procedure for digital data is largely the same as in a literature search where the reader is looking for published material on a particular topic. The search for both digital data and literature sources is assisted when the thematic or substantive characteristics of the items stored have been classified and indexed in some way.

Potential users of data from secondary sources are usually not only interested in their thematic content but also in two key geographical attributes: the part of the Earth's surface to which the data relate and the scale of the spatial units at which the data are recorded. These two characteristics are further ways in which a dataset may be classified and indexed. In some respects the first of these attributes concerns the population or universe covered by a dataset, while the second relates to the unit of observation within this population which has been sampled or measured. However, this is to take a rather superficial view of the nature of geographical data for two reasons. In some situations a user may wish to access a subset of spatial units (observations) in a geographical dataset, for example the grid squares within an area defined by two pairs of co-ordinates representing bottom left and top right of a rectangle, or those lying within a given radius of a point. In some cases the user may also wish to re-aggregate data to different spatial units to those contained in the 'parent' dataset, for example from 1 km^2 to 5 km^2 grid squares. The spatial indexing of geographical data has to anticipate a user's wish to subset datasets in a statistically legitimate fashion. It is therefore necessary for the indexing of geographical datasets to include details of the spatial referencing system employed and the nature of the spatial units of observation. For instance, a user searching for soils data on a particular region in Great Britain or a state in Australia (e.g. East Anglia and New South Wales respectively) needs to be sure that a classificatory category or search term such as Great Britain or Australia will identify the region and state as included within, or covered by, national datasets for smaller areas in those countries.

In the early years of accessing data from secondary sources, searches for thematically and geographically relevant data usually involved looking at dataset descriptions in published catalogues and directories, or alternatively having privileged information about surveys and data that were being collected by private companies or government departments. The prospective user's search for a suitable dataset, particularly when this 'holy grail' has yet to be lodged with a data archive, is still likely to entail examination of printed material. With the advent of the internet and the development of 'search engines' such as AskJeeves or Yahoo! for using the WWW the means of finding out about the existence of datasets is changing. However, many such WWW searches will only lead users to summary, 'pre-digested' information about a topic rather than to raw data for use in their own analysis, although this situation is changing.

3.5.2 Locational Metadata

Locational metadata involves two components: first, where data are stored in a computer system or archive; and, secondly, where data are held. The first is mainly a 'housekeeping' question which concerns data suppliers, especially as the number of datasets expands and it becomes necessary to know where copies of data files are stored, in what format and in what state of availability and processing. The determination of where datasets are held is of more direct concern to potential users. The search procedure for suitable datasets is likely to vary depending upon whether those organizations responsible for data collection are to be included alongside more accessible data archives, repositories and commercial data suppliers. In the former case, identifying whether a suitable dataset exists is potentially a time-consuming and exasperating process, not unlike searching for the proverbial needle in a haystack. Data archives and data suppliers are usually concerned not only with preserving data but also with making them available to users and therefore they tend to promote and advertise their wares in ways that bring them to the attention of potential users. Such promotional activity has occurred in various ways and has adapted to technological developments, with less emphasis placed on personal contact and more on indirect systems.

The WWW system has allowed these new types of data information system to be developed. Instead of providing full catalogue descriptions of datasets, the services may constitute a simpler directory which points users to people and organizations from whom more detailed information may be obtained. The WWW system itself, which permits users to 'leap' from one host site to another, then enables the user to make such enquiries of the person listed or search for more specific information on that organization's WWW pages. One early example was the Spatial INformation Enquiry Service (SINES) system. This was established in the UK in the mid-1990s by a consortium of government departments and was maintained by the Ordnance Survey. SINES was a prototype information service on geographical data generated and held by the British government and public agencies. The system enabled users to search information by reference to the thematic content and the spatial units referenced in a dataset. The development of such metadata information systems should enhance the potential users' ability to identify datasets which fulfil their needs, although clearly their value is dependent upon the maintenance of an accurate and complete directory.

3.5.3 Technical Metadata

A user's concern for the thematic content and geographical characteristics of datasets is closely related to such matters as the coding of attributes and measurement of variables, but, when the intention is to access raw data, the user also needs to know about the dataset's file and record structure, data item types and software format. Some of these are amenable to inclusion in the data descriptions already mentioned, others are only appropriate for those users seriously considering or actually accessing a particular dataset. The structure of geographical datasets is of particular concern to data archives and data suppliers, who need to exert some measure of quality control on the information they provide. The structure and format are also especially important when using datasets which were generated some time ago, even if this was only in the recent past, and are held in a way that is associated with earlier generations of software.

Technical information about the layout of the data records and the way in which variables have been coded or measured are clearly specific to an individual dataset. Until recently they were often held in printed documentation associated with a dataset, and might include a questionnaire schedule, a coding frame, a copy of the file used to read in the data and a technical description of data collection method, for instance giving details of sampling strategy. Early attempts to hold this information in a machine-readable format, for example as text files, were fraught with difficulty on account of the diversity and potential complexity of the information. The comprehensiveness and quality of this type of information depends upon the willingness of the data collector and the secondary source organization to provide and store it in a user-friendly format. Much of this technical detail has been obscured from, or become unnecessary to, the 'casual' data user as on-line information services have developed and the facility for interchanging data between different software and hardware systems has spread through common file types.

PART 2
CHARACTERISTICS AND REVIEW OF SECONDARY SOURCES

Chapter 4

Remotely Sensed Data

4.1 CHARACTERISTICS OF REMOTELY SENSED DATA

Remote sensing is defined as 'the observation of, or gathering information about, a target by a device separated from it by some distance' (Cracknell and Hayes, 1993: 1). In other words, it is a way of collecting information about a phenomenon without direct contact, and, by implication, usually without the target's knowledge. In the present context remote sensing refers to the collection of data relating to the Earth's surface environment. The temperature of oceanic water bodies could be determined by direct sensing using thermometers distributed at a series of sampling points (weather stations) or by means of processing digital satellite images which have recorded electromagnetic radiation (EMR) in the thermal infrared waveband. Most people are equipped with their own personal remote sensing devices, their eyes, which are able to detect EMR in the visible wavebands, and a personal processing facility, their brain, which enables the remotely sensed data to be converted into information in the form of visual images. The absence of direct contact with the object under observation does not remove the possibility of inaccuracy nor even of subjectivity in the collection of data. However, the advantages of such systems are generally regarded as the speed and regularity with which they can collect large quantities of data at a fraction of the cost of alternative procedures, notably field-based survey techniques. The definition of remote sensing given above embraces its use in a broad range of disciplines and non-geographic applications, and naturally the focus here is on its role with respect to collecting data relating to the environment and the Earth's surface.

This relatively simplified description of remote sensing illustrates the principles of the various techniques as a means of collecting geographical data, in other words, data referenced to and about features proximate to or comprising the Earth's surface. Attempts have been made to obtain data about the environment through remote sensing for nearly 150 years, when the first experimental photographs of the surface of the Earth were taken by balloonists over the village of Petit Biêtre near Paris in

1859. Little serious attention was paid to such exploits until aircraft were used for photographic military reconnaissance during World War I. By the mid-1920s and 1930s, aerial photography had emerged as a realistic proposition for research applications and for obtaining information for mapping purposes.

Receiving further impetus from its use for military reconnaissance during World War II, aerial photographic remote sensing grew in importance until it was partially eclipsed by the launch of the first orbiting satellite system in 1960. There have been considerable advances in the number and sophistication of satellite systems since the 1960s, although recently there has been a resurgence of interest in aerial photography. Table 4.1 provides a chronology of the development of Earth observation remote sensing systems mounted on aerial and space platforms since the mid-1800s. By including recent and planned satellite launches, it clearly shows how such systems are contributing to the current proliferation of geographical data.

The year 1960 therefore constitutes a recognized watershed in the history of remote sensing (Colwell, 1983; Cracknell and Hayes, 1993). Prior to 1960 many developments in remote sensing, a term not in general use at the time, were concentrated on military applications. Aerial photography was developing for use in civilian fields, notably in those concerned with monitoring land resource activities, such as agriculture and forestry. Images of the Earth produced by satellite-borne remote sensing systems in the 1960s emphasized, for both scientists and the general public, that 'The Earth was . . . an entity and its larger surface features were rendered visible in a way that captivated people's imagination' (Cracknell and Hayes, 1993: 4). The new remote sensing technology of satellites appeared as a valuable tool in providing the data that could be used to solve many of the Earth's natural and resource problems. Grandiose schemes were proposed and, in some cases, attempted to model global systems for the purposes of resource management and environmental prediction. Perhaps some over-selling of a new technology is inevitable, but early promise gave way to a realization of the significant difficulties and problems involved in interpreting the information gathered by satellite systems. By the 1990s a more sanguine approach had emerged, which recognized that remote sensing from satellites sits alongside other tools for collecting information about features on the Earth's surface. When these tools are used in conjunction with each other they can help to enhance our knowledge and understanding of environmental and human problems.

The main focus of this chapter is on the sources of remotely sensed data that gather information about the surface of the Earth, namely aerial photography and satellite-borne systems. However, it is important to recognize that there are other ways in which to transmit information about objects apart from EMR, such as through ultrasonic waves, and other ways in which to derive and interpret information about terrestrial features apart from in the form of images, for example through radio sounding methods. Gathering information about the features and events on the Earth's surface makes use of a wide range of instruments and platforms, which centre on the principle of detecting radiation at various wavelengths in order to produce images of the surface. Aerial photographs are referred to as continuous images and hold their data in an analogue form, which corresponds to and visually portrays human interpretation and perception of the phenomena contained in the image. In contrast, most remotely sensed images obtained from satellites are of a type referred to as discrete with the data held in digital as opposed to analogue form. Sections 4.2 and 4.3 respectively examine aerial photography and satellite remote sensing systems, while the remainder of this section will examine in more detail the general principles of image detection using EMR and the common features of geographical data obtained from these two types of remote sensing system.

4.1.1 Electromagnetic Radiation

The key to an environmental remote sensing system is the detection of the EMR reflected by terrestrial features or by the Earth's surface itself. The main source of EMR is the Sun with other sources, such as power stations and the Earth's emitted heat, of comparatively minor significance. All forms of EMR (e.g. X-rays, the visible spectrum and radio waves) travel at the speed of light in a wavelike fashion and create the electromagnetic spectrum, in which the wavelength of the radiation forms a scale or spectrum for recording the phenomenon of EMR. This spectrum is divided into a series of ranges according to specified wavelengths, for example the ultraviolet section from 0.3 to 0.4 μm and the visible from 0.4 to 0.7 μm (Figure 4.1). Only part of the EMR arriving at the outer limit of the Earth's atmosphere will reach the terrestrial surface. A proportion of the incoming EMR from the Sun in selected wavelength ranges (e.g. 1.4 μm and 1.8 μm) is intercepted by clouds, ozone and other matter in the atmosphere by means of absorption, reflection or scattering, a process known as attenuation. However, some EMR will reach the Earth's surface and be reflected back. This can be sensed by detectors directed towards the Earth that are mounted on aircraft, satellites or spacecraft. These are known as passive remote sensing devices, since the instrument simply receives whatever radiation arrives from the surface, recording quantities in one or more wavelengths. In contrast in an active detection system, the instrument transmits radiation towards a target, receives back the reflected radiation and records the amount in selected wavelengths.

The usefulness of EMR from a remote sensing perspective arises from the fact that individual objects emit or reflect EMR in different quantities along the electromagnetic spectrum, producing a characteristic series of peaks and troughs, which is referred to as the object's spectral response. The term spectral signature has also been used, although some writers have taken issue with this claiming that it 'gives a misleading impression of constancy and uniqueness' (Mather, 1987: 25), a comment which might of course also be made of a human signature. The nature of a feature on the Earth's surface, or of the surface itself, can therefore be inferred by processing the digital data received by the detection instruments. For example, the EMR reflectance, absorbency and transmittance pattern of healthy leaves is characterized by:

- low reflectance of blue and red light (0.4 μm, 0.6 μm)
- medium reflectance of green light (0.5 μm)
- high reflectance of near infrared (0.75–1.35 μm)

Table 4.1 Chronology of remote sensing

Date	Instrument	Country
1855	Hot air balloon	France
1860	Hot air balloon	USA
1898	Cameras attached to birds	France
1909	Aerial photography from aircraft	Italy
1915	Aerial camera	
1914–18	Aerial photography from aircraft	World War I countries
1918–39	Aerial photography from aircraft Development of stereoscopy	USA and Canada
1930–40	Development of radar	Germany, UK and USA
1939–45	Aerial photography with infrared film	USA
1950s	Development of space observation systems commences	USA and USSR
1957	Sputnik satellite	USSR
1960	TIROS I (NOAA and ESSA)	USA
1960	Use of near, middle and thermal infrared film	USA
1969	Apollo 9	USA
1972	Landsat 1 (ERTS programme commences)	USA
1973	Skylab	USA
1975–78	Landsats 2 and 3	USA
1978	Seasat	USA
1980	METEOR	USSR
1981	Space shuttle	USA
1982–84	Landsats 4 and 5	USA
1983	Kosmos I	USSR
1984	Kosmos II	USA
	NOAA 9	USSR
1986	SPOT I	France
	NOAA10	USA
1987	MOS	Japan
	IRS – LISS1 and LISS2	India
1988	NOAA 11	USA
1990	SPOT 2	France
	MOS 1b	Japan
1991	NOAA 12	USA
	ERS-1	Europe (ESA)
1992	SPOT Asia	Singapore
	JERS	Japan
1993	SPOT 3	France
	PoSat	Portugal
1994	Atlas	USA
	NOAA-GOES (8)	USA
	ERS 2	ESA
	FY-2	China
	GMS 5	Japan
	GOMS	Russia
1995	NOAA-GOES 1	USA
	Resours 02	Russia
	CBERS	China–Brazil

Table 4.1 (*cont.*)

Date	Instrument	Country
	IRS 1-C	India
	Radarsat	Canada
	ERS-2	Europe (ESA)
1996	CBERS	China–Brazil
	ADEOS	Japan
	Almaz 2	Russia
	Earth Watch	USA
	Radarsat	Canada
1997	Earth Watch	USA
	NOAA-GOES II	USA
	OrbView	USA
	Space Imaging	USA
1998	SPOT 4	France
	ENVISAT	Europe (ESA)
	Landsat 7	USA
	EOS AM 1	USA–Japan
	EOS Color	USA
	KOMSAT	Korea
	Space Imaging	USA
	SAGE III/Meteor 3M-1	USA
1999	IRS 1-D	India
	SPOT 5A	France
	ADEOS II	Japan
2000	EOS-PM1	USA
	POEM/Envisat 1	European Organization for the Exploitation of Meteorological Satellites (EUMETSAT)
	METOP	Europe (ESA), USA, European Organization for the Exploitation of Meteorological Satellites (EUMETSAT)
2002	*EOS-ALT 1*	*USA*
	EOS-CHEMI 1	*USA*
2003	*EOS-AM 2*	*USA*
	EOS-AERO 2	*USA*
2004	*EOS-AM 2/L-8*	*USA*
	SPOT 5B	*France*

Italics denotes planned launches.

Sources: Campbell, 1987; Harris, 1988; Chuvieco, 1990; Kramer, 1994; Treadwell, 1995; Asrar, G. and Greenstone, R. 1995; Aranha, 1998.

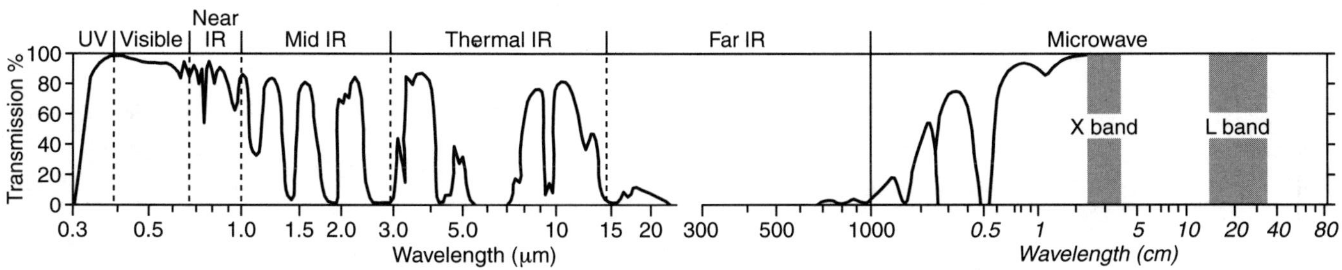

Figure 4.1 Waveband ranges on the electromagnetic spectrum.

The Earth's atmosphere does not intercept EMR to a uniform extent along the full range of the electromagnetic spectrum, which results in certain groups of wavelengths (wavebands) being favoured for the purpose of remote sensing:

- visible or optical (wavelength 0.4–0.7 μm)
- near (reflected) infrared (wavelength 0.7–3.0 μm)
- thermal infrared (wavelength 3–14 μm)
- microwave (wavelength 5–500 mm)

This characteristic enables remote sensing devices to be tuned to receive and detect emitted EMR in one or more of these ranges, in a similar fashion to a radio 'picking up' the sound waves of a broadcasting station transmitting on a particular wavelength.

The EMR sensors are usually mounted on a platform (e.g. a satellite or aircraft) and fall into one of two categories: first, those which instantaneously capture an image of the scene by using a lens to focus light from the field of view onto a flat photosensitive target, such as a film emulsion in a camera. The second has a detector which traverses over the target in a series of parallel lines building up a complete image from a series of cells or scenes. The former produces analogue images (aerial photography or video footage) whereas the latter results in digital images. A digital image comprises a number of cells, known as pixels or picture elements, each of which has a discrete intensity value that is proportional to the EMR received by the sensor. Analogue images can easily be converted into digital versions, for example by means of scanning, which may prove useful for classification purposes, although current developments in digital photography and aerial videography are likely to reduce the need for the intermediate stage of an analogue image. The numerical values in a digital image can be processed by an analytical technique known as image analysis in order to identify the nature of the features occurring on the Earth's surface.

General issues relating to the measurement characteristics of geographical data discussed in previous chapters also apply with respect to remotely sensed data. EMR, as a variable to be measured, can be regarded as multidimensional in the sense that its intensity in different regions of the spectrum can be quantified simultaneously for a given unit of the Earth's surface. Analogue and digital sensors can be designed to record an image in either a single spectral band or in a series of spectral bands (a multispectral sensor), respectively exemplified by an infrared camera and Landsat's Thematic Mapper. Multispectral sensors are more useful in environmental applications of remote sensing because they record the intensity of EMR in a specific group of waveband ranges from each pixel, which improves the analyst's ability to match the reflectance pattern to the established spectral responses of different terrestrial features. The situation is similar to building up the socio-economic character of an area by taking a single variable, such as the number or percentage of people in single year age groups. High percentages of people below 5 years and between 20 and 30 years might indicate areas where young families were prevalent. In practice, the socio-economic characterization of areas uses several attributes, each of which contributes to the overall character of the area, in order to generate a composite indicator. In the case of remote sensing, supplementary data will be used to aid with the image analysis and interpretation, and specifically to train the classification system. In both cases a form of 'data triangulation' is used to determine in one case the environmental and in the other the socio-economic characteristics of places on the Earth's surface.

Unlike the methods of measurement or enumeration of a target using direct contact methods covered in the next three chapters, remote sensing does not directly generate either the location or the attributes of geographical phenomena. These fundamental characteristics have to be derived by interpreting and processing the remotely sensed data into a form so that they are capable of representing features in a recognizable fashion and in their correct orientation on the Earth's surface. Even the analogue images produced by aerial photography may give misleading information about the nature, shape and size of the phenomena they portray unless some form of interpretation is employed. Buildings may, for example, appear to lean in certain directions or their shape may be distorted.

4.2 AERIAL PHOTOGRAPHY

Aerial photographs were first generated over 100 years ago, but they only became important as a source of environmental information following World War I. This section aims to review the characteristics of aerial photography and to introduce the techniques of their interpretation for the novice user. The intention is not to examine in detail the analytic techniques of photogrammetry, which are adequately covered in other texts (Curran, 1981; Warner *et al*, 1996; Lillesand and Kiefer, 2000), other than when this is necessary for comparing different types of aerial photography and for understanding their usage. The intention is to enable current or potential users of aerial photographs to be alert to the questions they should be asking and the issues they should be considering when setting out to use this source of geographical data. In some situations it is possible for people seeking to obtain aerial photographs for an area to commission such a survey for their specific purpose. In these circumstances there is more opportunity for the end-user to control some of the factors which affect the photographs that are produced. However, once such missions have been undertaken and the commissioning organization has completed its analysis, other researchers may be able to use the same photographs for other projects. This outline of aerial photography as a source of georeferenced data will act as an introduction to the review of sources for remotely sensed data in Chapter 8.

The equipment required to produce simple aerial photographs is relatively basic – a light-proof box, a lens, and a film coated with an emulsion sensitive to EMR – and a means of traversing this equipment over the Earth's surface at altitude in a relatively controlled fashion. Inevitably these basic requirements are available in more or less sophisticated versions, which leads to variations in the quality and therefore the usefulness of the photographs produced. The earliest experiments with aerial photography used balloons and even cameras strapped to pigeons' breasts in order to take photographs from the air. Nowadays aircraft and satellites are used to raise the camera over the Earth's surface or into space, with the result that all of the Earth has now been photographed. The main concern here is not with large format photographs taken from space, of the type taken by the first satellites in the 1960s, but with small format photographs. Such aerial photographs have been collected for most of the Earth's land surface and are relatively cheap in comparison to field surveys and satellite images. They provide a synoptic view (i.e. of a wide area at the same time and under the same conditions) and can thus be used to examine spatial and temporal variability at a regional or local scale. A further advantage of aerial photography is that overlapping pairs of photographs can be taken, which provide a three-dimensional perspective when viewed stereoscopically.

In the context of this book, aerial photographs are self-evidently not maps, otherwise they would be covered in the next chapter. However, they are also 'not maps' for a more important reason: the objects depicted on a map share a common scale, whereas features represented on a photograph have different scales. The origin of the difference between aerial photographs and maps lies in the fact that a photo is a product of perspective or central projection, and a map is the product of orthographic projection (Warner *et al*, 1996: 3). Thus two rectangular objects on a map with the same dimensions implies that the real-world phenomena which these represent also have the same surface area. Two seemingly identically sized objects on an aerial photograph in planar view may have different surface areas because the angle at which they have been captured by the EMR detector (camera) may not be constant across the entire photograph. Figure 4.2 makes this point by portraying an aerial photograph of a regular square grid with the camera positioned vertically over the central square. The squares nearer to the edges become progressively more distorted either vertically or horizontally, while those towards the corners are deformed in both planes. A regular grid of this type is unlikely to occur naturally on the Earth's surface. However, some of the squares in the grid might conceivably be the roofs of identical

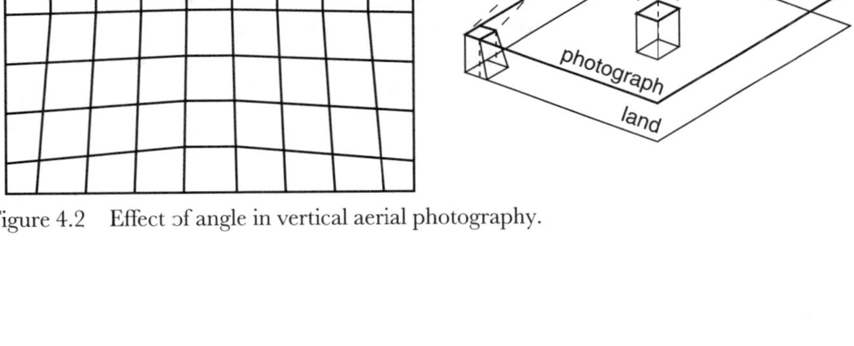

Figure 4.2 Effect of angle in vertical aerial photography.

skyscrapers, in which case the aerial photograph might suggest that their cross-sectional shape and area was different. It is important to make this distinction, since, when viewing an aerial photograph and observing its seemingly accurate registration of the geometrical characteristics and geographical location of terrestrial features, it is easy to be under the misapprehension that these can simply be transferred from photographic to cartographic media. If the Earth's surface was entirely flat and the platform (aircraft) on which the camera was mounted maintained a constant altitude and orientation then this might be possible; however, these three conditions rarely pertain. Aerial photographs in fact have two types of scale: the average scale and the point scale. The former refers to the overall mean scale for a particular aerial photography mission or for an individual photograph. The term point scale refers to a photograph's scale at a point on the ground at a specific elevation.

Despite the difference between the meaning of scale in respect of maps and photographs, aerial photography has frequently been used as a source of information about features on the Earth's surface in order to produce maps. There is therefore a need to translate between the multi-scaled environment of an individual aerial photograph or a collection of photographs and the single (common) scale of a map. As with most forms of data collection the process of converting the raw data into a format that is capable of useful interpretation involves consideration of a series of parameters relating to how the data (photograph) was collected. In the case of aerial photography these are geometrical parameters referring to the focal length of the camera, the displacement or distortion of the target object, the co-ordinate axes of the photograph and the tilt of the aircraft in relation to the centre of the photograph. The process of converting between the two media – aerial photographs and maps – is known as photogrammetry, which involves the geometrical integration of these parameters.

The principle underlying stereoscopic aerial photographs is that successive pairs of overlapping frames, by changing the point of observation of a target feature, provide a three-dimensional image. Our eyes, which are a fixed distance apart, also work on the principle of stereoscopy to view objects from two perspectives simultaneously. This provides us with depth of vision and the ability to assess distances and the relative position of objects. The overlapped portion of two adjacent photographs along a flight line can be viewed stereoscopically with each of the two photographs in a pair being analogous to looking at an object firstly with one eye and then the other. When looking at an object in this way it appears to move relative to its background: viewed with the left eye the object appears to shift to the right; to the left when viewed with the right eye; and holds a median position when both eyes are used. The displacement of an object is known as parallax and the extent to which the object appears to move is related to its distance from the points under observation and the base width between a person's eyes. The idea is the same with an aerial photographic camera, except that rather than simply two eyes at a fixed distance apart, the camera progresses along a flight line taking separate but overlapping, regularly spaced photographs of the surface beneath. Features on the overlapping portions appear in two adjacent photographs and the elevations of features can be determined by means of photogrammetric manipulation. The ratio of the distance between the points of observation (B) and the height of the camera (H) is an important quantity since it provides a way of assessing the accuracy of the height measurements between terrain features. The larger the B:H ratio the greater the accuracy of the height measurement.

Aerial photographs can be classified in different ways, according to the characteristics of the camera, the type of film and the position of the camera relative to the horizontal. The most common type of camera used for aerial photography is the single lens reflex (SLR), which enables the target to be sighted through the lens. These cameras are available in 35 mm and 70 mm small and medium formats, which refers to size and shape of the negatives produced (70 mm negatives are square whereas 35 mm are rectangular). Large format cameras (23 cm) are standard for mapping purposes. The standard lens for 35 mm SLR cameras has a focal length of 50 mm, with the equivalent focal length of 80 mm for 70 mm format cameras. Focal length refers to the distance between film plane and the centre of the lens when focused on infinity. Figure 4.3 illustrates how focal length directly influences the camera's field of view, which in turn affects the scale of the photograph. Increasing the focal length of a camera mounted at a constant altitude lengthens the distance on the photograph between two points on the ground. SLR cameras are often designed so that interchangeable lenses can be fitted, with those having a focal length longer than the standard referred to as telephoto and those shorter as wide angle. A wide angle lens with a 28 mm focal length on a camera mounted at a given altitude will have a 74° angle of view in comparison to a 46° for a standard 50 mm lens. SLR cameras can also be fitted with zoom lenses enabling the focal length to be varied. These are not recommended for use in light aircraft aerial photography since they can produce greater distortion, are bulky and have a limited light-gathering capacity.

The four groups of film used for small format aerial photography are panchromatic (black and white), near infrared black and white, colour and false colour near infrared (Curran, 1981). This classification refers to the wavelength bands of EMR to which the film emulsion is sensitive when exposed by opening the camera's aperture. Panchromatic film is widely used because of its geometrical stability, low cost of processing and printing, and high spatial resolution. Black and white near infrared film has the advantage of sensitivity at near infrared wavelengths which makes it useful for detecting crop diseases and soil moisture status. Colour film is readily available, easy to process and enables the human eye to distinguish more features because our eyes are more sensitive to tints and shades of colour than tones of grey. False colour near infrared has a wide range of applications, for example in detecting tree stress in forests and identifying crop marks on archaeological sites, on account of the incorporation of near infrared with some visible wavelengths of EMR. In all cases the films are available in 35 mm and 70 mm formats, although larger formats tend to be used in professional applications.

The position of the film plane in the airborne camera, whether fixed to the aircraft or hand held, in relation to the horizontal determines if the aerial photographs are vertical or oblique. Vertical photographs are usually regarded as the ideal form, since if truly vertical they possess no distortion or displacement arising from the tilting of

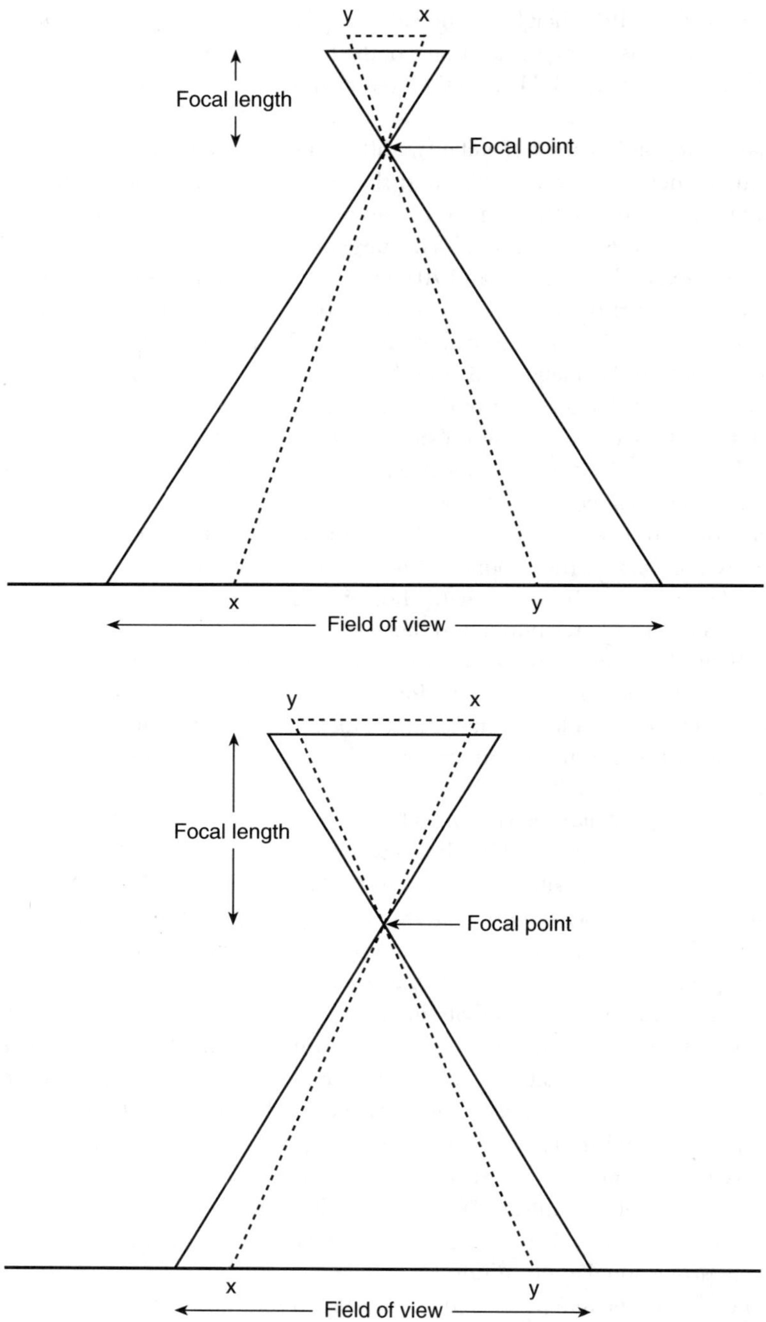

Figure 4.3 Effect of field of view in aerial photography.

the aircraft, and will only incorporate such imperfections through topographic variation. Operational conditions often mean such perfect verticality cannot be consistently achieved throughout a mission and a tolerance level of less than 5° for the film plane and aircraft from the horizontal is generally accepted as vertical. Above this angle the photographs are referred to as oblique, which can be further classified as low or high depending on whether the horizon is included within the camera's field of view, in which case it is called a high oblique.

The increased usage of information technology for handling geographical data has had two important consequences for aerial photography: first, the traditional equipment used for photogrammetry has been replaced by digital photogrammetric workstations; and secondly, the transfer of detail from printed photographic media to paper maps has been superseded by the digitizing and scanning of aerial photographs. Consideration of the second of these changes provides a link to Section 4.3 which concentrates on the digital images generated by satellite remote sensing systems.

Digitizing from aerial photographs involves transforming locational information from a photograph's co-ordinate system to ground co-ordinates using a minimum of three, but preferably more, ground control points. The control points are at readily identifiable locations on the ground which have known Cartesian co-ordinates, for example Universal Transverse Mercator x, y, z values, derived either by field survey using a Global Positioning System (GPS) or digitized from an existing map. The conversion procedure in the case of a single photograph is known as space resection and uses collinearity equations to match the co-ordinates of the control points expressed in terms of the x,y axes of the photograph to the known Cartesian co-ordinates of the same points on the ground. The orientation of the photograph's x,y co-ordinate system is determined by its alignment along the flight line of the aircraft (x axis) and by the y axis being perpendicular to this. The z axis of the photograph becomes important if the landscape of the area covered has relief. In such circumstances, a Digital Elevation Model (DEM) is required, which is a regular grid of x, y, z co-ordinates that acts as a model for the surface terrain, and hence is sometimes known as a Digital Terrain Model (DTM). The DEM is used to generate the elevation, z, of a given x,y defined point. The operations can be carried out using digital monoplotters which allow individual photographs to be converted into a digital form for analysis and integration with other types of geographical data in a GIS.

Stereoscopic digital aerial photographic mapping is inevitably more complex and requires more sophisticated equipment than the process just outlined. Nevertheless, the essential principle of the procedure is the same, although it is complicated by the need to match the co-ordinate systems of two photographs with equivalent x, y, z ground co-ordinates. One complicating factor arises from the need to take account of what is known as y parallax, which is the displacement of an object measured along the y axis in the overlap portions of two adjacent photographs. This displacement originates from differences between the two photographs in tilt and/or flying height of the aircraft at the time of film exposure. Stereoplotters are divided into analogue and analytical types. Analogue plotters are based on mechanical systems, although they are able to interface with computers. Analytical plotters rely on producing a mathematical solution using digitized information from the photographs. Most of the analytical stereoplotters on the market perform three key functions (Warner *et al*, 1996: 57): orientation of the photographs, computation of ground co-ordinates from photo-measurements and display of information graphically or digitally. One of the most significant aspects of digital photogrammetric instruments is their ability to output information for mapping and cartographic modelling within the more widely available GIS software.

Scanning of aerial photographs, which registers each point (pixel) on an image with digital values denoting its position and brightness, represents an alternative way of achieving analogue to digital conversion. As with most computer equipment, whether for use in graphics or more general applications, there are various types of scanner which are differentiated in terms of the quality of results they produce. Scanners are either simple hand-held or flatbed systems which use a charge coupled device (CCD) sensor to pass over the illuminated original photograph or alternatively move the photograph in relation to the static sensor. Two differentiating characteristics of scanners are their pixel resolution, usually measured in terms of dots per inch (dpi), and their radiometric resolution, normally expressed as the number of bits stored. Dpi signifies the spatial dimensions of the image resulting from the scanning operation and therefore can be directly related to the size of the photograph being scanned. A typical 23 cm by 23 cm aerial photograph (9.0625 ins by 9.0625 ins) scanned at a pixel resolution of 4800 dpi would result in a digital image of approximately 394 220 pixels ($9.0625 \times 9.0625 \times 4800$). The radiometric resolution can most easily be thought of as the density or intensity of brightness for each pixel recorded as a value on either a panchromatic or colour scale. On the one hand, this produces a scale going from white through varying degrees of grey to black and, on the other, a series of colours through the visible spectrum from blue to red. The number of grey scale or colour shades which can be recorded by a scanner relates to the bits (binary digits) of digital information stored. An 8 bit scanner can store 256 colours.

4.3 SATELLITE IMAGERY

Satellites and other spacecraft may carry cameras which will take pictures of the Earth from space, but it is more common for analysts to use the digital rather than analogue form of image which can be collected by the sensors on these platforms. Digital images produced by sensors mounted on satellites or other spacecraft have become increasingly important since the 1960s as a source of remotely sensed data. The twofold purpose of this section is to review the various satellite systems which have been in operation over this period and to introduce the essential elements of image processing and interpretation. Other texts concerned solely with remote sensing and digital image manipulation have covered image processing in more detail than here (Curran, 1985; Mather, 1987; Lillesand and Kiefer, 2000). The remainder of this section will provide the details necessary for users to understand the principles of satellite sensing systems and the essential elements of how image processing can turn the digital data they collect into interpretable representations of features at the Earth's surface.

4.3.1 Satellite Systems

Several satellites have been launched for environmental remote sensing since the first Television and InfraRed Observation Satellite (TIROS-1) in April 1960 and analysts therefore have to select carefully from the available imagery those parts that fulfil their particular purposes (Allen, 1984). Initially the main emphasis was on collecting data for meteorological purposes and for modelling global systems, but subsequently remote sensing for environmental analysis has become important. There are three main characteristics that distinguish between different satellite systems: the nature and speed of the satellite's orbit or relative position to the Earth; the spatial resolution and viewing area; the spectral resolution in conjunction with the range of intensity values recorded. These sources of differentiation will be discussed in general terms before comparing the characteristics of particular satellites.

The nature and speed of the satellite's orbit refers to its relationship with the Earth and the planet's rotation within the solar system. The two main types of remote sensing satellite are geostationary and polar orbiting. Geostationary satellites occupy a fixed position in relation to a particular area of view on the Earth's surface and are positioned over the equator – they are stationary with reference to the Earth (Figure 4.4a). There is a network of geostationary satellites collecting data for meteorological purposes at approximately 60° to 70° intervals around the Earth that are operated by the USA, Europe, Russia and Japan. Polar orbiting satellites move in an approximately circular orbit around the Earth passing over the poles (Figure 4.4b). The altitude and speed of a satellite's orbit in relation to the Earth's rotation directly influences the frequency or repeat rate of viewing for a given area of the Earth's surface. These parameters define the temporal resolution in a remote sensing system. The sensors scan parallel lines across the Earth's surface perpendicular to the satellite's orbit (Figure 4.4b). These lines or swaths are of a defined, constant width and the Earth's rotation shifts successive orbit paths westward so that during a relatively fixed period of time the whole surface of the globe is covered. Some satellites have a short temporal resolution and record information for the same area every 12 hours whereas others are longer at up to 16 days. The speed of orbiting satellites is important because of the need in most cases to be Sun-synchronous, in other words to view an area of the Earth which is receiving EMR from the Sun and reflecting it back to the sensor. Ideally, the time of day when a particular area of the Earth's surface is in the view of the sensor should be the same for each passage of the satellite. Temporal resolution is also relevant in relation to the probability of a satellite viewing an area on a cloudy day, the more frequently images are collected for a given area the greater is the likelihood of recording data when there are cloud-free skies.

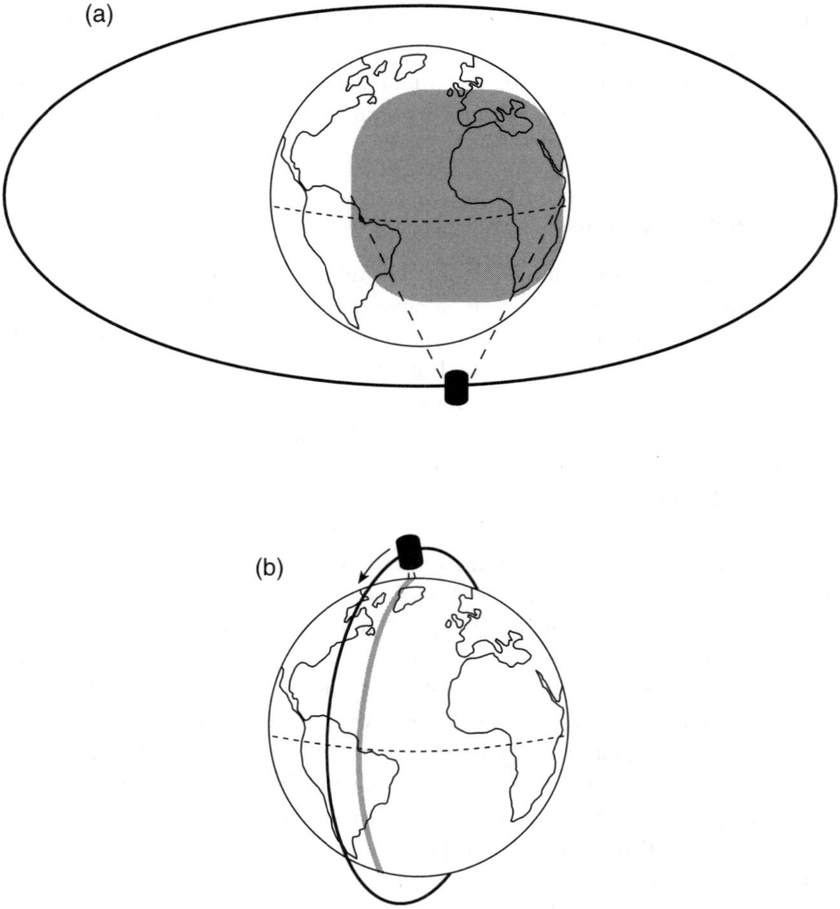

Figure 4.4 Orbits of remote sensing satellites. (a) Geostationary orbit, (b) polar orbiting.

Furthermore, when environmental monitoring and the operation of physical processes are under investigation, the frequency of data collection needs to be carefully selected to correspond to the rate of change in environmental features.

Spatial resolution in remote sensing is more difficult to define than temporal resolution and 'can be measured in a number of different ways, depending on the user's purpose' (Mather, 1987: 36). Initially spatial resolution might seem easy to define in the same way as with aerial photography as the distance between two image points that can be resolved with the human eye (Warner *et al*, 1996: 111). Unfortunately, this definition is inapplicable to remote sensing digital images in their raw, unprocessed form, since they themselves are not immediately resolvable by the human eye, but are a matrix of numerical values recording intensity of EMR wavelengths. Intuitively the spatial resolution of a digital imaging system might be thought of as the area on the ground viewed by the detection instrument at a given altitude and point in time. Assuming that there is no framing system in place to rectangularize the image, this area would be circular. This provides us with a commonly-used definition of spatial resolution known as the Instantaneous Field of View (IFOV). The diameter of the circle forming the IFOV for a particular satellite can vary because of fluctuations in the altitude of the geostationary position or orbit. Another reason for being cautious about using IFOV as a geometrical definition of spatial resolution is that does not take into account the spectral properties of the target (Mather, 1987: 39). Mather provides a more comprehensive discussion of alternative measures of spatial resolution, including Effective Resolving Element and the Effective Instantaneous Field of View, than is appropriate here. The main advantage of these alternative definitions is that they incorporate the system's ability to detect and record the radiance of targets at a number of discrete points and to enable the accurate identification of terrestrial features irrespective of their relative size and radiance properties.

It is important not to confuse the spatial resolution of an imaging system with pixel size. A pixel is a single cell in a digital image which corresponds to a specified rectangular or square area on the ground surface and it has values associated with the radiance from this area according to the one or more wavebands that the sensor is able to detect. The division of the Earth's surface into a grid of pixels in this way is of course entirely arbitrary. The pixel grid in a digital image can be thought of as capable of floating over the Earth's surface with the radiance values for each pixel adjusting according to variations in radiance. Suppose that the right side of a single pixel was precisely aligned along the edge of the concrete roof of a building, assuming the concrete was uniform, the pixel would have constant EMR across its extent. If the pixel was to 'float' towards the right so that it spanned the concrete of the roof and the adjacent healthy vegetation, then the digital record would become confused, reflecting the different EMR spectral signatures of the two types of land cover. As the pixel size is changed, for example, through re-sampling of data, so the radiance values and the ability to identify terrestrial features improves or diminishes.

However defined, the spatial resolution of satellite systems has progressively reduced in size since the 1960s, with some digital images now spatially resolved to 1 m by 1 m units on the Earth's surface. This reduction in spatial resolution should not be taken to imply that applications of satellite remote sensing should always strive to use images with the smallest possible resolution and disregard those with a larger size as out of date or old fashioned. As in all instances where secondary sources of data are used it is important to select from the range of available sources those data which are most appropriate to the purpose in hand. A detailed spatial resolution may be advantageous if the phenomena under investigation are relatively small. In contrast, a spatial resolution 5 km by 5 km in size may be entirely suitable in some global applications. The decision for the investigator is similar to selecting an appropriate number of points at which to measure altitude, either in the field or by interpolation from contours on a topographic map, when constructing a DEM. Absolute and relative variations in height will be less significant across an area of relatively low relief than in a mountainous area. A regular 100 m grid of points in the former case would certainly produce a large number of redundant height measurements where adjacent points are at the same altitude; in the latter instance the grid may be too coarse to capture relief changes. In the case of a digital image the measurements are not recorded at the intersections of the grid lines (grid points), but are a mean value of EMR for the entire grid square. If the geographical phenomena under investigation is relatively undifferentiated across large regions of small grid squares, there will be data redundancy and it would have been entirely feasible to have used a digital image with a larger spatial resolution. When using digital images it is important to be aware of whether the intention is to identify the presence or absence of some terrestrial target feature, for example, the crowns of trees or the roofs of houses, or alternatively to quantify a variable (attribute), such as a soil temperature or leaf moisture content.

The spectral resolution of a satellite is defined by the number and size of the wavebands to which its sensor(s) is (are) tuned . Given than EMR is measured along a scale (see Figure 4.1), it is possible to segment it into many different wavebands and for combinations of these to be detected by a particular satellite remote sensing system. Most satellite systems are multi-spectral, which means that they have sensors collecting data for more than one discrete band width. Multi-spectrally scanned digital images are more useful than those derived from a single spectral band because they enable the interplay between the radiance of several wavelengths captured simultaneously from the Earth's surface to be utilized when processing the data in order to detect features. We have already seen that the so-called spectral signature of terrestrial features or variables provides a means of distinguishing between one feature or variable value and another. It would therefore seem reasonable to suggest that having a sensor with a very finely tuned, high spectral resolution, which closely corresponded with the known spectral signature of the target feature (e.g. granite), would improved the sensor's success and accuracy. Unfortunately, the penalty for high spectral resolution is a reduction in the signal-to-noise ratio.

Most scientific measuring instruments achieve their results with a certain, quantifiable level of accuracy and satellite sensors are no exception. This noise is the unwanted variation in the signal received by the sensor which arises from mechanical or electronic imperfections in the equipment. Increasing the sensor's bandwidth enlarges the range of signals which are accepted by the sensor. The notion is similar to widening the percentage range of marks corresponding to a particular class of degree in order to reduce the proportion of borderline candidates. Some modern types of satellite sensor use a pushbroom system which lowers the signal-to-noise ratio by using a bank of individual sensors in a row, each of which views a single scan line element across the swath. This is opposed to systems in which the view of a single detector is electromechanically shifted to each scan line element in a swath in turn.

Digital images from satellites can record EMR at different levels of intensity or spectral resolution in a similar way to the scanning of aerial photographs. As indicated previously this concept relates to the number of grey scale or colour shades, which is usually stated in terms of bits which the sensor can record. For example a 4 bit sensor can store 16 levels of grey scale or colour shades represented by data with integers from 0 to 15, while with an 8 bit system this increases to 256 levels with values from 0 to 255. There is an interaction between the number of levels which can be stored and the signal-to-noise ratio, since the incremental change from one level to the next must be sufficient to correspond at least to more than the noise level. Although higher spectral resolution is likely to produce images which are visually more aesthetically pleasing to the human eye, it should not be assumed that using 256 levels as opposed for example to 64 will necessarily lead to a greater ability to distinguish between features (Tucker, 1979).

Since the first Earth observation remote sensing satellite was launched by rocket in 1960, numerous sensors have been launched in this way or using other forms of spacecraft (e.g. Space Shuttle). Most of the individual systems have been part of programmes, in which successive satellites in a series were designed to replace redundant units once they reached the end of their working life and/or improve on the system's capabilities in respect of some or all of the characteristics which have just been discussed. Some of the main satellite programmes which have produced useful data for scientific and engineering purposes are:

- GOES-E, GOES-W and METEOSAT
- LANDSAT-1, LANDSAT-2 and LANDSAT-3
- TIROS-N, NOAA-N
- LANDSAT-4, LANDSAT-5, SPOT, MOS and ERS-1

Table 4.2 provides generic information about the satellites in a series by means of thumbnail sketches and indicates some of the principal applications to which their data are suited. One of these systems will now be examined in more detail to reveal the evolution of satellite remote sensing.

The French government commenced development of the *Système Pour l'Observation de la Terre* (SPOT) in 1978 with SPOT-1 launched in 1986 and the most recent (SPOT-5) in 1999. Satellites in the series are Sun-synchronous polar orbiting at an altitude of 832 km with a repeat cycle of 26 days, although the innovative system of redirecting the sensor's optics to view either side of the perpendicular swath path enables more frequent scenes of any individual point on the Earth's surface. The first three satellites in the series carried high resolution visible (HRV) imaging systems operating in either 10 m resolution panchromatic mode over the EMR range 0.51–0.73 μm or 20 m resolution multi-spectral (colour IR) mode in the ranges 0.50–0.59 μm, 0.61–0.68 μm and 0.79–0.89 μm. Data are transmitted to ground receiving stations within a radius of 2600 km with onboard magnetic tape storage acting as backup and for download when out of range of a receiving station. The fourth SPOT satellite extended the series' previous EMR range by including a sensing capability in the mid-infrared range (1.58–1.75 μm) with a ground resolution of 20 m. A vegetation instrument was also added to the payload of SPOT-4 linked to the high resolution visible and near infrared system (HRVIR), which provides a 2250 km swath with a spectral resolution at nadir (perpendicularly beneath the satellite) of 1 km. The latest SPOT satellite (5) has been designed to replace the

Table 4.2 Characteristics of selected satellite systems

Satellite sensor	Image repeat frequency	Sensor type	Spectral resolution	Spatial resolution (m)	Swath width (km)	Main applications
Meteosat						
Meteosat 1	30 min	Radiometer	Visible	2400	Dull Earth or regions	Meteorology, vegetation and environmental
Meteosat 2	30 min		Medium IR	5000		
			Thermal IR	5000		
TIROS-N/NOAA						
	12 h	Advanced Very High Resolution Radiometer (AVHRR)	Visible and near IR, middle IR, 2 thermal IR bands	1100	2400	Meteorology, oceanography, hydrology, and vegetation
Landsat						
Landsat 1	18 days	Multi-spectral Scanner (MSS)	Six bands – 4 visible, 1 near IR and 1 thermal IR	79	185	Land use, vegetation, geology, geomorphology and hydrology
Landsat 2	18 days			237 (thermal)		
Landsat 3	18 days					
Landsat 4	16 days		Seven bands – 3 visible, 1 near IR, 2 middle IR and 1 thermal IR	30		
Landsat 5	16 days			120 (thermal)		
Landsat 7	16 days	Thematic Mapper (TM)			185	
SPOT HRV						
MS	26 days	Multi-spectral Scanner (MSS)	Three bands – 2 visible, 1 near IR	20	60	Land use, vegetation and mineral exploration
P		Panchromatic (P)		10	60	
Space Shuttle						
MOMS	Limited cover	Multi-spectral Scanner (MSS)	Three bands – 1 visible, 1 near IR and 1 thermal IR	10–20	140 (Stereo)	Land use, vegetation geology and hydrology

SPOT-4 HRVIR with two high resolution geometric instruments, which provide a higher spatial resolution respectively 5 m for panchromatic and 10 m for colour mode. Further, a high resolution stereoscopic instrument provides fore-and-aft data for the creation of a global 10 m digital elevation model.

4.3.2 Overview of Digital Image Processing

A digital image comprises a matrix of pixels containing digital numbers, which represent the proportion of EMR at a particular wavelength. In the case of multi-spectral images, the matrix is multi-layered with each layer containing data for the same pixels in a different spectral range (e.g. LANDSAT-5's TM collects data in three bands). These may be images captured at one point in time or multi-temporal, where images of the same area have been obtained for different time periods (e.g. images of agricultural land taken during the land preparation and growing seasons). Digital image processing may be thought of as a group of analytic techniques which enable the digital information to be translated into a form whereby the user can accurately identify and interpret the Earth's surface and the features thereon. Throughout the development of image processing, this type of analysis has been associated with relatively specialist hardware and software systems. The principal reasons for this are the large quantities of data associated with digital image data files and the complex multivariate statistical operations which may need to be performed on these data. Despite the general changes in the character of information technology hardware and software outlined in earlier chapters, image processing systems retain the need for a certain amount of specialist equipment for digital data input, processing, storage and output. Image processing may be thought of as the transformation between a digital view of the Earth's surface, including its three-dimensional component, and a view in which information technology can be used to help with interpreting and distinguishing features or for measuring certain variables. Optical interpretation of aerial photographs has an essentially similar purpose, but is applied to analogue or non-digital images. However, if such images are converted into digital files through scanning, then some image processing procedures can be applied.

The hardware component of an image processing system has become increasingly transparent to the user, although the requirement for a means of transferring, storing and processing large volumes of data and the facility to display output on a high resolution screen (usually 1024 by 1024 pixels) and a similar quality plotter or printer, establishes the need for a fairly high specification workstation or PC system. Most image processing software guides the analyst by means of default options through the normal or typical choices that may be made with respect of the selection of processing operations to be carried out on a particular type of image. Clearly, when the purpose of the image analysis is driven more by scientific research and development rather than by practical application and the search for solutions, such routinizing of procedures may be overridden. The five main operations carried out in an image analysis system are: pre-processing, enhancement, transform, filtering and classification. A full account of these operations is beyond the scope of this book and the following overview is intended to summarize the stages involved in manipulation of digital image data from the numerical format into a classified image in order that novice users of digital imagery may comprehend the ways in which the raw data may have been adjusted. More detailed discussion can be found in Mather (1987), Curran (1985) and similar specialist remote sensing and image processing texts. Throughout these data processing operations, it is important not to lose sight of the fact that their essential purpose is to lead to the accurate interpretation of the terrestrial features and characteristics recorded by the satellite-mounted sensor. Image interpretation constitutes a further final stage in which an assessment of the quality of the analysis is undertaken. A geographical examination of the results will also usually involve a ground truth survey. This is ideally collected at the same time as the remotely sensed image was gathered for a series of sample points, but this is often impossible because of cost and time constraints.

4.3.2.1 *Pre-processing*

Most geographical data obtained from secondary sources require some measure of prior treatment in order to eliminate anomalies and identifiable errors before they can be used for reliable analysis. The problem is to decide which tasks should be undertaken, since this will vary according to the nature of the application. Some will be necessary because of the particular source of digital imagery being used (e.g. sixth-line banding effect on some LANDSAT images), but would not otherwise be needed, whereas others would invariably be carried out (e.g. geometrical correction), but may occasionally be unnecessary. Pre-processing operations for satellite imagery may therefore entail one or more of the following: removal of any defects or errors; geometrical transformation of the image co-ordinate system; and the suppression of effects caused by atmospheric scatter.

Two important data deficiencies may be present in a digital image: first, a missing scan line is indicated by either a black or a white stripe; and secondly, a regular striped pattern can signify a systematically repeating defect associated with a defective array of detectors. Pre-processing operations seek to overcome these problems by replacing missing pixel values or adjusting values to compensate for the striping effect. The geometrical correction or registration procedures applied to digital images involves similar issues and has a comparable aim to the pre-processing operations undertaken on aerial photographs. Digital images from remote sensing satellites are not maps and do not in themselves have accurate geo-references yet the environmental information which they contain frequently needs to be combined with maps. Geometric correction fits the scale and projection of digital images to a standard co-ordinate system (e.g. Universal Transverse Mercator) and thus enables the classified image to be integrated with cartographic representations on the same projection by using a GIS. Various kinds of geometric error can be corrected including distortion arising from the viewing angle of the sensor and skewing of the image occurring through the velocity of the Earth's rotation, which varies relative to the satellite's latitude. Registration is concerned with ensuring that two or more images of the same area but taken at different times share a common co-ordinate system, which, depending on the particular application, may or may not be tied in with a standard map co-ordinate and projection system. Some of the radiant energy in the Earth's atmosphere is scattered and absorbed (attenuated). The signal received by a sensor as it views a particular portion of the surface will partly be composed of EMR emitted from the features on or the characteristics of that surface and partly of EMR scattered within the atmosphere or from other locations on the surface. The sensor cannot discriminate between radiance which is truly reflected from the pixel in view and the component arising from the scattering.

4.3.2.2 *Image Enhancement*

Image enhancement covers procedures which have the purpose of making the digital image more easily interpreted by the human eye. Different methods may be used depending upon whether the data are recorded as mono- or multi-spectral images. In the former case the two main forms of data manipulation comprise procedures for contrast stretching, and black and white to colour conversion. Most of the adjustments are carried out by means of lookup tables that define an association between the input pixel values and the output display. This enables changes to be made to the former without the values in the data file being permanently altered, since such a change would inhibit further analysis and processing. In the case of multi-spectral images, one of the main methods of enhancement involves carrying out multivariate statistical operations on the pixel values by means of principal components analysis, although it is possible that some of the following procedures may be applied to individual bands in a multi-spectral image. Image enhancement operations for monochrome satellite imagery may therefore entail adjusting the contrast between pixel values and/or display as pseudo-colour images.

On some occasions images will be detected in which the spread of pixel values is very narrow, in other words its visual contrast is small. In the case of a panchromatic image, it may appear excessively dark or light which makes it difficult to interpret the presence of terrestrial features. Altering the contrast of an image involves applying some adjustment to a histogram of pixel values. Figure 4.5a presents a histogram of pixel values with a grey scale range of 1 to 16, which would produce a relatively dark image lacking in contrast, since the large majority of pixels occupy the range 4 to 8. The full output intensity range (1–16) can be used by assigning values 0 through 34 to 0 and 126 through 255 to 255 and stretching the input pixel values in a linear fashion across the range 35 to 125 (Figure 4.5b). Alternatively, histogram equalization can be applied, which seeks to produce output intensity values with an approximately equal number of pixels but recognizing that individual pixels with the same input value should not be separated. Another approach to the problem involves transforming the distribution in input pixel values to correspond with the Gaussian or normal frequency distribution (Figure 4.5c).

Conversion of a black and white image to colour produces a pseudo-colour display, so-called because the pixel values captured in the one-dimensional grey scale still do

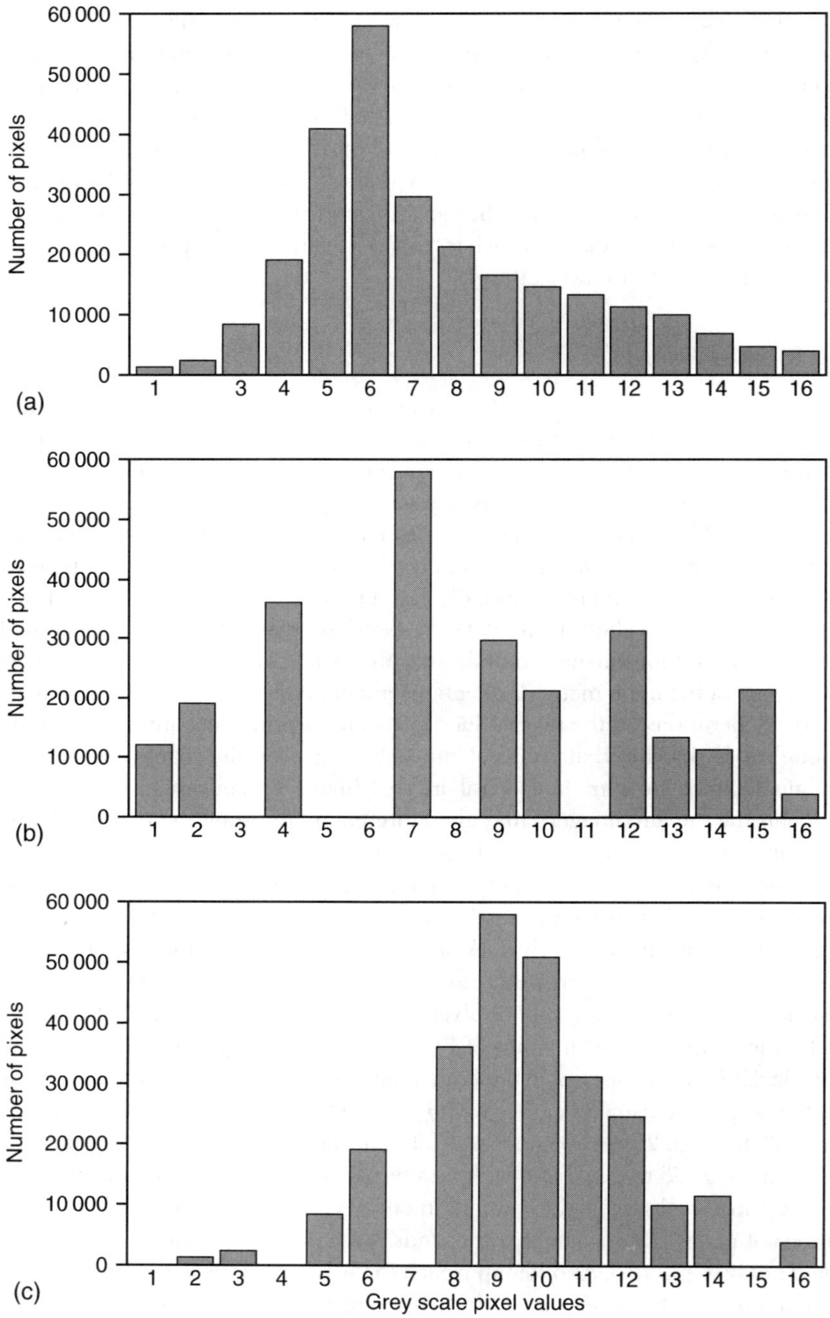

Figure 4.5 Contrast enhancement of images. (a) Skewed distribution of pixel values, (b) linear stretch, (c) Gaussian stretch.

not contain true information in the red, green and blue bands, as would be the case in a colour image. Three methods are commonly used to produce pseudo-colour images: density slicing; pseudo-colour transform and user-defined colour transform. Density slicing is achieved by assigning contiguous grey scale values to a point in the RGB colour cube. Pseudo-colour and user-defined colour transforms are similar in that both entail assigning a discrete colour to each grey scale value. They are differentiated simply by the fact that the latter allows the user to specify the format of the transformation.

4.3.2.3 *Image Transform*

The term 'transform' in respect of digital imagery refers to a collection of arithmetic operations that create a new composite image from two or more bands of a multi-spectral or multi-temporal image. The purpose in the first case may be to determine characteristics of the vegetation cover for the image and in the second to quantify change in soil moisture deficit. The types of image transform that may need to be carried out are strongly related to the aims of the analysis being undertaken and to the fitness of the particular imagery for achieving these. Image transform operations for satellite imagery may therefore entail one or more of the following: simple arithmetic combination and multivariate analysis.

It is important when carrying out arithmetic operations (addition, subtraction, multiplication or division) on different images or separate bands of the same image that they are identically registered using the same co-ordinate system, otherwise there will be a mismatch in the relationship between the pixels and the Earth's surface in the different images. Figure 4.6 illustrates how these arithmetic operations can be carried out in a simplified representation of an image. The group of pixels in the top left section of Figure 4.6 have been added and re-scaled (sum of input values is divided by number of images) to retain the same dynamic range in the output. The pixels in the top right of the 'image' have been subtracted, which may be useful when processing images captured at different times, for example to detect change in vegetation zones. Multiplication may be used in order to isolate and display only selected parts of an image; this can be achieved by having a false, binary image where the pixel values are 1s and 0s, which will mask out unwanted features from the true image. Differences between spectral bands are often detected by means of division or ratioing. Some pixels in the lower right part of Figure 4.6 are very similar to each other in the two input images and so when division takes place, a result near to 1.0 is achieved, in other cases the values for the same pixel are more dissimilar and produce results larger or smaller than 1.0.

Principal components analysis (PCA) is one of a group of multivariate statistical techniques known as factor analysis, which aim to collapse a large number of input variables into a smaller number of 'composite' variables known as components in the case of PCA. The original variables 'load' on each component having an eigenvalue of $\geqslant 1.0$ by means of a correlation coefficient and loadings closer to -1.0 and $+1.0$

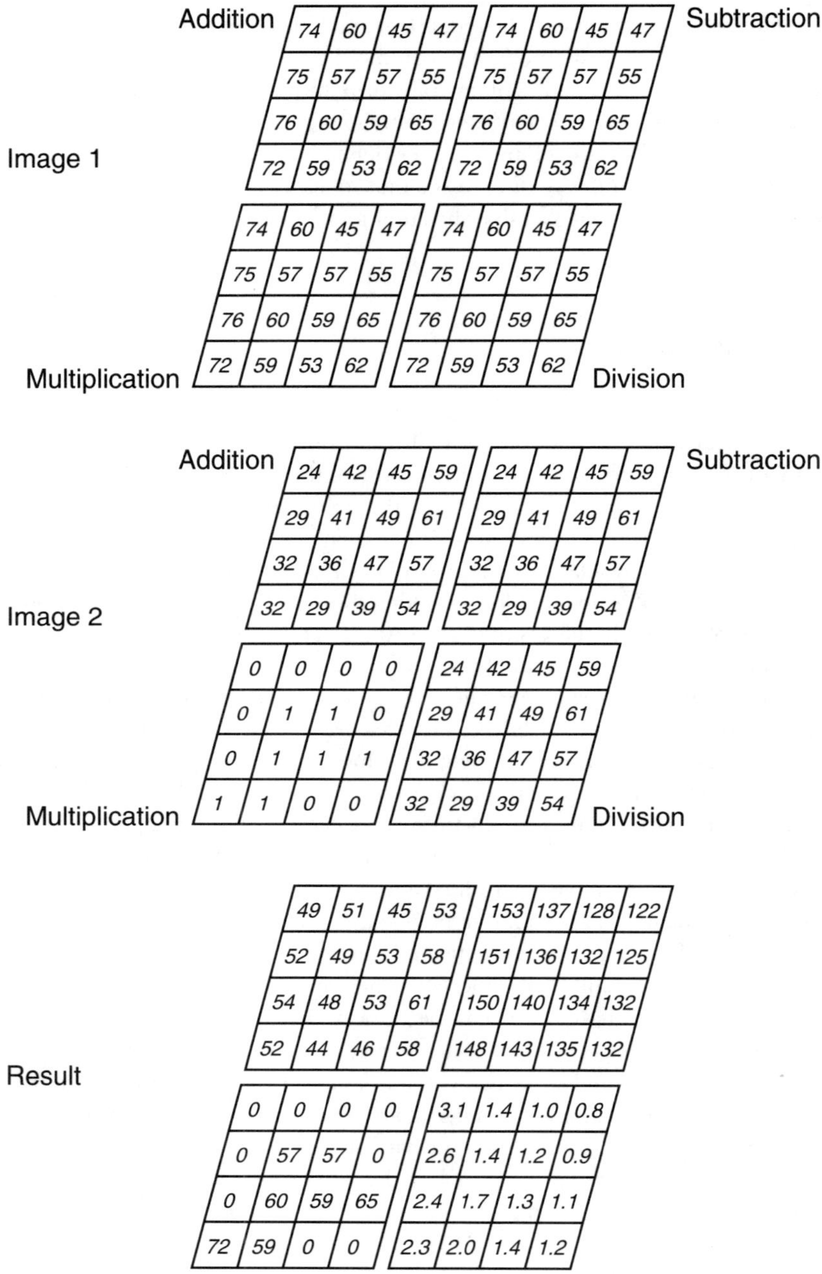

Figure 4.6 Arithmetic operations on image pixel values.

(e.g. <-0.6 or $>+0.6$) are used as a means of deciding which input variables should be used to define the component. The scores of the pixels in the image in respect to the components can be calculated and then displayed. PCA has been used with multi-temporal and multi-spectral images such as Landsat TM bands 1 to 7. In the latter case it is conceivable that the data will collapse down to one or two components with strong positive loading on most bands, which implies that there is considerable overlap in the information conveyed by each. This is of course not surprising since the underlying variable being measured (EMR) is the same in each case.

4.3.2.4 *Image Filtering*

The collection of pixels to which digital image data relate is an arbitrary framework superimposed on the Earth's surface for which either one or more 'average' EMR values are captured by the sensor. An individual pixel may not and, given its regular shape and arbitrary position is unlikely, to fit precisely with features on the Earth. Thus some pixels within a particular image fall entirely within a uniform feature parcel, whereas others will lie across boundary lines. Furthermore, the boundary lines of terrestrial features may not be as definite as we may imagine and in fact possess fuzzy boundaries. Figure 4.7 illustrates how an area of woodland vegetation may not be separated from adjacent grassland by a simple line, since the canopies of trees on the boundary often grow over the neighbouring grass. There is a difference depending upon whether the boundary is viewed in plan form from above or obliquely from the side. Any pixels lying over such a boundary are likely to have an indistinct spectral pattern. Image filtering techniques selectively emphasize or suppress information in the image in order that pixels provide a clearer digital message. The filtering techniques are essentially about trying to cope with either the fact that surface characteristics or features may have extensive variation over a short distance or may have minimal variation over a long distance. In other words the filters attempt to identify a scale component in an image. Filtering operations for satellite imagery may therefore entail one or more of the following: smoothing (low-pass) filters, sharpening (high-pass) filters; and edge detection.

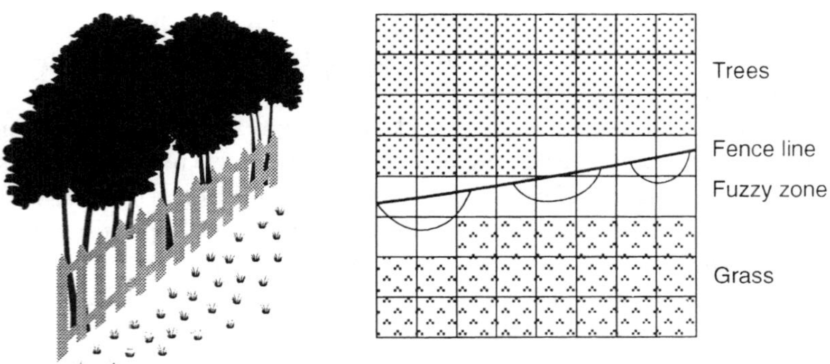

Figure 4.7 Example of fuzzy boundary between grass and tree vegetation cover.

Smoothing filters, as their name suggests, are concerned with smoothing out the highs and lows of pixel values either with respect to single or multi-band images. The process is analogous to adjusting for year-on-year variations in time series analysis. In the case of image processing smoothing attempts to ameliorate abrupt changes in value between adjacent or proximate pixels. A moving average three-point one-dimensional filter is illustrated in Figure 4.8 in a simplified form. A number of pixels (3) are taken in overlapping groups along vectors of the image, on each occasion the mean pixel value is calculated. The first becomes the value of pixel x_2 in the new image, the second x_3, the third x_4, and so on to x_{n-2}. A two-dimensional filter can also be used composed of a positive, odd but not necessarily equal number of pixels. Pixel values for the new image are derived by moving the filter window across the original image and calculating the mean value of the centre of a pixel (Figure 4.8). Similar one- and two-dimensional filters can be obtained by taking the median rather than the mean input value of the pixels.

Sharpening filters are essentially the opposite of low pass, smoothing filters and are used to remove blurring in an image. Although a relatively simple subtraction method exists, the more common filter is based on the mathematical concept of the derivative, which is the rate of change of a continuous function at a particular point. In simple graphical terms the derivative is the gradient of a curve, the extent by which the slope is increasing or decreasing at any point. The principle can be applied to images in order to determine the rate of change in pixel values, which occurs in the two-dimensional space of the rows and columns of an image. One important use of sharpening filters is to help with detecting edges, such as between land and water bodies or buildings and their surroundings.

4.3.2.5 *Image Classification*

Image classification assigns pixels to categories on the basis of their digital number in order to reduce the total range of digital numbers and spatial entities into the smallest number of meaningful categories. Once classification of an image has been carried out it may be possible to identify groups of pixels individually coded into a particular class, for example the vegetation category 'coniferous trees' that comprise a geographical unit, such as an area of coniferous forestry. However, because the data still exists in a rasterized format, there is no topological connection between these pixels within the database that 'informs' the system that they constitute a woodland polygon. This would require passing the data across to a geographical information system that was capable of raster to vector conversion and of creating the topological linkages. The classification of pixels is fundamentally a question of deciding how each pixel in an image should be labelled. This problem can be approached by either supervised or unsupervised classification techniques.

Supervised classification starts from the basis that the surface area of the Earth covered by an image contains a number of identifiable patterns and that the features comprising these patterns can be determined by means of statistical values. A series of templates can then be defined and applied to the pixels in an image in order to find

	A	B	C	D	E	F
1	**74**	60	62	55	53	45
2	**75**	57	55	51	56	57
3	**76**	60	63	56	57	59
4	72	59	48	56	57	53
5	54	72	51	65	67	75
6	59	71	52	60	65	79

74	60	62	55	53	45
75	57	55	51	56	57
76	60	63	56	57	59
72	59	48	56	57	53
54	72	51	65	67	75
59	71	52	60	65	79

74	60	62	55	53	45
75	57	55	51	56	57
76	60	63	56	57	59
72	59	48	56	57	53
54	72	51	65	67	75
59	71	52	60	65	79

74	60	62	55	53	45
75	57	55	51	56	57
76	60	63	56	57	59
72	59	48	56	57	53
54	72	51	65	67	75
59	71	52	60	65	79

75	59	60	54	55	54
74	59	55	54	57	56
67	64	54	59	60	62
62	67	50	60	63	69

74	**60**	**62**	55	53	45
75	**57**	**55**	51	56	57
76	**60**	**63**	56	57	59
72	59	48	56	57	53
54	72	51	65	67	75
59	71	52	60	65	79

74	60	62	55	53	45
75	**57**	**55**	51	56	57
76	**60**	**63**	56	57	59
72	**59**	**48**	56	57	53
54	72	51	65	67	75
59	71	52	60	65	79

74	60	62	55	53	45
75	57	55	51	56	57
76	**60**	**63**	56	57	59
72	**59**	**48**	56	57	53
54	**72**	**51**	65	67	75
59	71	52	60	65	79

74	60	62	55	53	45
75	57	55	51	56	57
76	60	63	56	57	59
72	**59**	**48**	56	57	53
54	**72**	**51**	65	67	75
59	**71**	**52**	60	65	79

	65	58	56	54	
	63	56	55	56	
	62	59	58	61	
	60	59	58	64	

Figure 4.8 One- and two-dimensional filters.

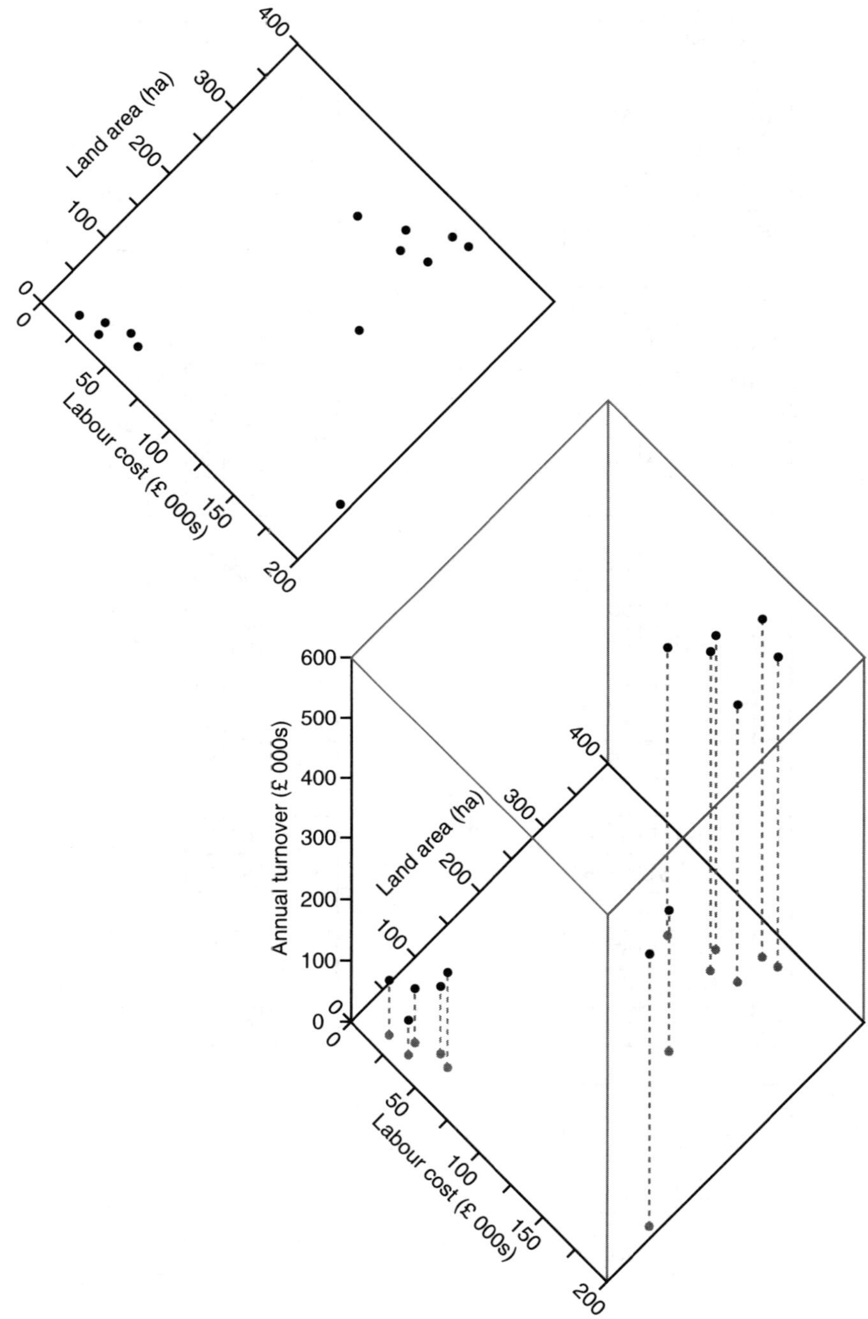

Figure 4.9 Two- and three-dimensional feature space.

places where there is a match between template and pixel pattern. The process uses field survey or ground truth information to train the system to classify the pixels in the digital image into the known categories. The procedure can be seen as analogous to how people cope with assimilating new information in their everyday lives: they try to put the phenomena they encounter into categories or classes based on prior knowledge. Hence, when new cohorts of students arrive at their university, they will often ask each other questions – What 'A' Level grades did you get? Did you have a gap year? What football team do you support?, etc. These simple pieces of information are then used to build up a stereotype of what the person is like. Unsupervised classification proceeds with no such *a priori* template, but uses statistical techniques that seek to identify patterns within the data (pixel values) themselves.

Most classification techniques used in image processing and other forms of classificatory analyses can be thought of as trying to identify clusters of observations within a multivariate feature space. The necessary existence of both spatial (locational) and attribute information in order to define and differentiate geographical features inevitably means that multivariate analytical techniques are required. However, it is difficult to visualize these geometrical situations graphically, although it is feasible to consider the simplified case of where features are located within two- or three-dimensional space (Figure 4.9). It is clear that in both cases the points locating the farms in respect of land area, annual turnover and expenditure on labour have patterns that enable groups or clusters to be identified. Multivariate classification techniques quantify this visual process of pattern recognition by defining groups based on distance between the phenomena. Applying this procedure to the multiband EMR measurements for a collection of pixels can assign them to clusters and thus enable feature patterns to be identified.

Chapter 5

Cartographic Data

5.1 CHARACTERISTICS OF CARTOGRAPHIC DATA

Maps essentially have two general functions: to present or visualize geographical data; and to archive or store geographical information. Both functions have been important since ancient times because maps provide a way for societies to communicate knowledge about the world around them and to preserve this information for future generations. Formal definitions of maps abound and often refer to a map as a graphic model of the spatial aspects of reality (Kraak and Ormeling, 1996: 42); or as a representation or abstraction of geographical reality. Maps are tools for presenting geographical information in a way that is visual, digital or tactile (Board, 1990); or are graphic representations on a plane surface of the Earth's surface or part of it, showing its geographical features. These are positioned according to pre-established geodetic control, grids, projections and scales (Steigler, 1978). The second definition implies that a map is something physical or tangible, which is clearly not the case with the category of maps known as 'mental maps'. These are the personal abstractions or visualizations which people hold in their minds of the world around them and will vary in detail and extent according to a range of factors, including education, experience and age (Gould and White, 1974). Whatever the merits of these definitions, maps are produced for use in many different situations and appear in many different guises (see, for example, Dorling, 1995). However, regardless of their precise form, the underlying aim of a map is to convey information visually in an accessible and convenient fashion having due regard to the users' perception of and ability to interpret spatial information. In many cases, maps will be produced by one person or organization for use by other people, which emphasizes their importance as a tool of spatial communication.

With the onset of GIS, maps took on an important new function, they were not only passive sources and media for archival storage of spatial information, but also active and dynamic tools for answering geographical questions, for modelling spatial processes and undertaking analysis. During the first 25 years of the development of

GIS, there was a need to source data relating to the location of basic geographical features and to convert these by means of digitizing or scanning into a format which could be used in the rapidly developing technology. Existing, usually printed maps constituted a major source of geographical data for these computer-based systems. This is not the place to resurrect earlier debates about the difference between computer-aided or digital cartography and GIS (Rhind, 1977; Waugh and Taylor, 1977). However, for the purposes of clarity and consistency in this chapter, the former is regarded as concerned with using computers to produce or revise maps mainly from primary sources of data obtained by survey, whereas GIS concentrates on the analysis, interrogation and manipulation of cartographic representations of geographical features and their associated attribute data. In practice the distinction between computer-aided cartography and GIS has become blurred since the debate raged over two decades ago, which is partly because the makers of maps have increasingly taken advantage of the developing information technology associated with GIS and partly because the maintenance of a distinction became increasingly futile.

However, there is an important distinction between cartographic data (i.e. maps) and some of the other sources of geographical data covered in other chapters, when maps are used as an input to a GIS. Consider the difference between a geographical analyst using an aerial photograph or remotely sensed digital image and a thematic map as a source of information about land use. A set of parameters about the aerial photograph will be known including the particular date and time it was taken and the average scale of the photograph (see Chapter 4). The map should also include a similar set of details including the date when the field survey was carried out and the map's scale. So what is the difference? The difference is that the user of the map is presented with a series of pre-defined land use classes, as depicted by the shading and/or colours in the legend or key, whereas analysis of the raw data from the aerial photograph or satellite image can be tailored to produce various land use classifications suiting particular purposes. The geographical data on the map has already been processed before reaching the GIS analyst, whereas the aerial photograph awaits this treatment. Maps are nevertheless important sources of topographic data for GIS analysis, indeed in some instances they may be the only available source when it is impractical to undertake primary data collection. However, the availability and use of global positioning systems for determining location on the Earth's surface increasingly obviates the need to use existing maps for even this basic type of information.

Maps, of course, are not the only sources of geographical data which will usually have experienced some form of pre-processing before reaching the GIS analyst or end-user. Raw data from an agricultural census are likely to have been coded and aggregated to geographical units, remotely sensed data may have undergone some form of compression and decompression, and administrative data will undoubtedly have been anonymized by the removal of the subjects' names, addresses and other identifiers. Users of maps as sources of geographical data for GIS analysis should appreciate that they will always have been compiled or generated from at least one other, intermediate data source. As a result of this the opportunity for the user to amend any hidden analysis of the raw data, which the map maker has carried out, will be limited. Maps, whether in analogue or digital form, on hard copy or electronic media, are usually produced from the processing of raw data collected by another means, including topographic field survey, photogrammetric processing of aerial photographs, satellite remote sensing and global positioning systems. Some maps are produced as a result of bringing together data collected by one of these means with attribute information derived from censuses and sample surveys, or administrative records. In parallel with other chapters in Part II, the focus here is on a generic source of geographical data for use in geographical interrogation and analysis, in this case maps in analogue or digital formats. The information contained on maps comes to the GIS user secondhand, since it will have been processed by another individual or organization. A map is thus not a primary source of data for use in a GIS, but a secondary or possibly even tertiary source. The increased remoteness of the end-user of cartographic data from the original primary source may exacerbate some of the generic problems of using geographical data through secondary sources.

The aerial photograph and land use map example above highlight a traditionally important division within the group of data sources referred to here as cartographic data: this division relates to topographic and thematic mapping. This distinction reflects two fundamentally different types of geographical data which may or may not be contained in the same finished map product. The topographic component of geographical data essentially relates to a spatial image of the Earth's surface (Kraak and Ormeling, 1996: 44), which shows where features are located in relation to one another and to the overall spatial referencing system. The thematic element concerns the information stored about, but often independently of, those spatial features. Before examining the characteristics of topographic and thematic mapping in Sections 5.2 and 5.3, there are some general issues relating to the characteristics of cartographic data which need to be considered.

5.1.1 Map Projections

Co-ordinate systems, as explained in Chapter 1, are one of the most important ways of georeferencing spatial data. The basis of such systems is a regular grid of equal sized square units, which can be referenced at different levels of specificity. Such Cartesian grids are entirely satisfactory for relatively small areas of the Earth's surface, but are unsuited to larger areas because of the near spherical nature of the Earth. The global georeferencing system identifies a location on the Earth by means of latitude and longitude, which are lines lying parallel respectively to the primary horizontal and vertical axes of the Earth (the Equator and Greenwich Meridian). A location is determined by reference to the angles between its meridian (latitudinal) plane and, first, the Equator and, second, east or west of the Greenwich Meridian (0°). The near spherical (ellipsoidal) nature of the Earth means that at different latitudes the physical distance between 1° varies. Most national mapping agencies have developed their own local grid networks which are tied trigonometrically into the global system. The British National Grid, which is not used in Ireland, was devised in 1930s. It comprises a grid of 91 100 km by 100 km squares, which can be further subdivided into 10 km, 1 km and 100 m squares (Figure 5.1). The 100 km squares are referred to by a two-digit letter code, for example ST relating to the 100 km square between eastings 300 and 400, and northings 100 and 200. The

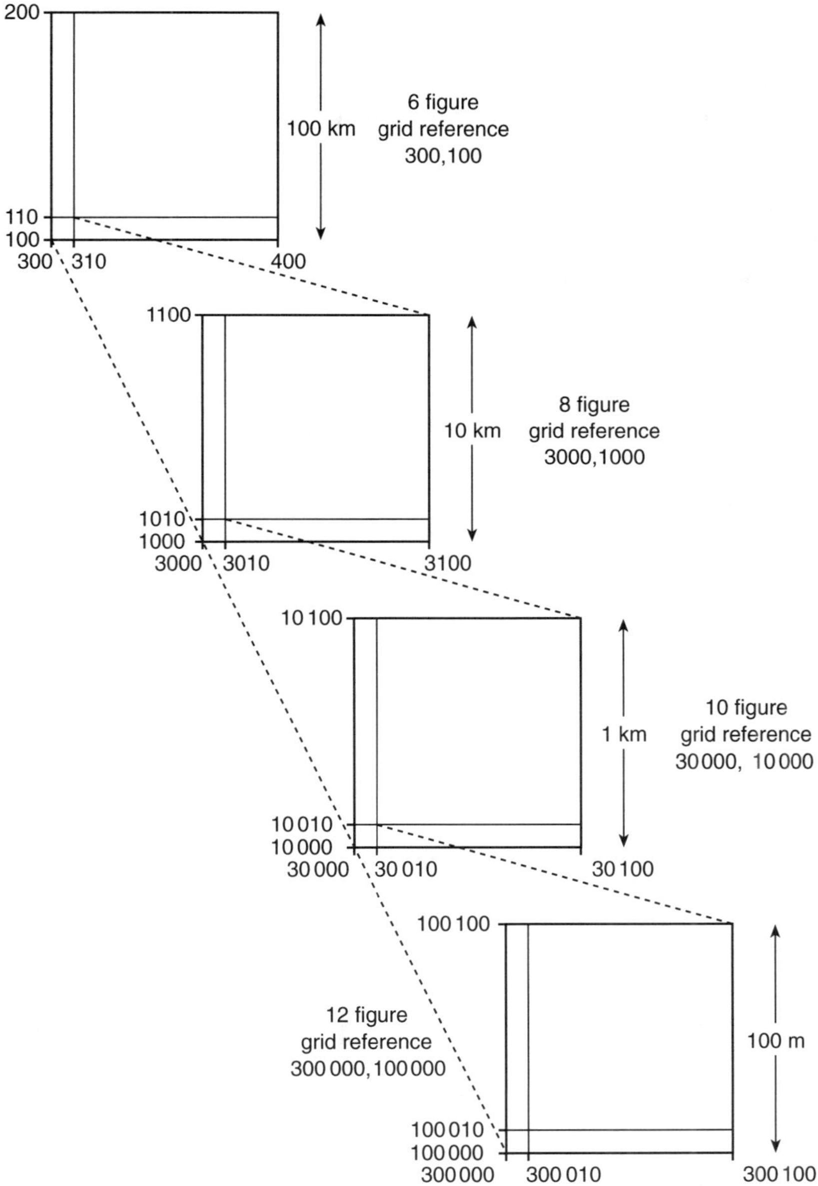

Figure 5.1 Georeferencing using the British National Grid.

two-letter code may be omitted and replaced by a full numerical co-ordinate reference. As the size of the grid squares decreases so the number of digits required to locate features in terms of grid co-ordinates increases, thus a four-digit reference is accurate to the nearest 1 km, whereas a six-digit reference has an accuracy of 10 m.

When mapping larger areas it is necessary to convert the co-ordinates of the global referencing system from the three-dimensional ellipsoid to a two-dimensional plane. This conversion process is achieved by mathematical transformation, which results in map projections categorized into three main groups: azimuthal, cylindrical and conical. Each of these can be further divided according to whether it is normal, transverse or oblique with respect to the Earth's north–south axis. Using a map projection inevitably produces some distortion of shape, area or distance between geographical features. This distortion provides a further way of classifying projections into those which are conformal (object's shape is preserved), equal-area (object's area is maintained) and equidistant (distance between objects is sustained). Figure 5.2 illustrates the three principal categories of map projection. When choosing a map projection for a particular application the nature of the distortion is the first guide (Kraak and Ormeling, 1996: 84). The main issue from the GIS analyst's perspective is to be able to determine which projection has been employed when using cartographic data that have been produced by another organization, particularly when the analysis requires maps from different sources to be used. It will be apparent from the previous discussion that the importance of this question when planning GIS analysis is linked to the size of the study area. If the area is relatively small and/or the source maps have the same projection, then the question is of little significance. In contrast if global or continental maps (e.g. for Canada or the USA) are to be used or produced then map projection becomes more important. Fortunately modern GIS software is capable of automatically transforming between co-ordinate systems thus enabling information originally mapped according to different projections to be merged. However, assuming the choice of projection was made in a rigorous fashion when the diverse maps were originally compiled, it should of course be noted that any subsequent transformation and harmonization might produce unexpectedly distorted results.

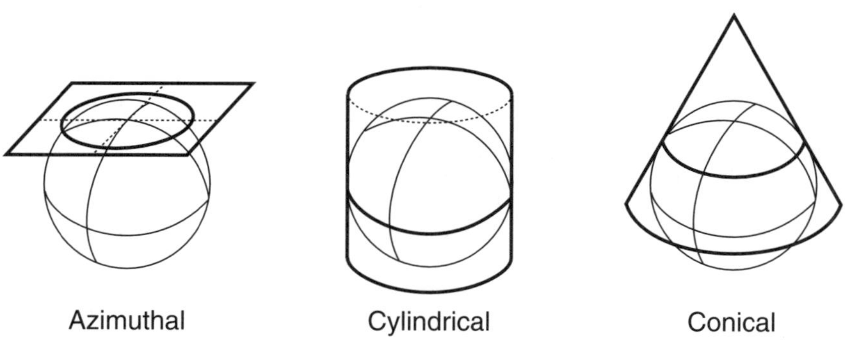

Figure 5.2 Commonly used map projections.

5.1.2 Data Capture and Conversion

When maps have been used as sources of data for GIS analysis in the past they were almost invariably only available in a hard copy, printed format. This impediment has almost disappeared in many developed countries, as organizations producing maps increasingly use computer technology for the purpose. The conversion of the data contained on a map into a digital or machine readable form prior to GIS analysis may be less of an obstacle than was formerly the case. Consequently the principles involved in the conversion of data held on a map into a digital form will only be outlined here, since they have been discussed extensively in numerous GIS texts (Burrough, 1986; Star and Estes, 1990; Tomlin, 1990; Berhardsen, 1992; Heywood *et al*, 1998). There are two issues worth highlighting relating to the conversion of cartographic data into a digital format (digitization): first, the conceptual representation of geographical features; and secondly, the practical procedure for capturing the data. The information contained on a map is an abstraction or model of the real world: features are included or excluded; attributes are categorized and shown selectively; map design employs conventions. Conversion from this analogue, printed abstraction into a digital format posed the fundamental problem of devising a model which would represent the perceptual, physical and phenomenological structure of two-, three- or four-dimensional real-world geographical features in a digital form. The two main approaches for representing geographical features in such a format are the raster and vector models. Differentiation between these systems remains important in spite of technological developments having brought them closer together and having achieved the integration of data held in the two systems. In essence, they represent different ways of viewing the structure of phenomena occurring on or near the Earth's surface: one view being that these vary continuously over space according to a mathematical function; and the other that space is populated by discrete objects with clear, irrefutable boundaries.

The raster model, following the first of these views, uses standard spatial units, usually squares in a regular grid, as an intermediate means of specifying those parts of the Earth's surface which go to make up or form part of geographical features. The position of each square in the grid is specified by means of row and column numbers, which can be converted into the co-ordinate systems employed in map projections. Figure 5.3 illustrates how natural and man-made features of the real world, in this case geology, land use and altitude, can be identified by means of a series of raster grid layers. Squares are assigned to a single type of feature (e.g. a wood when mapping areas of forestry or woodland or Lower Chalk in the case of solid geology) by means of an attribute code number or colour, here shown as a particular shading pattern. If more than one category of geographical attribute (e.g. land use and altitude) is to be stored in a GIS, it is necessary for each class to be held in a separate data layer, since each raster unit in a grid can only contain one item of data in addition to its row and column position. In practice, each point or spatial unit which is superimposed on the Earth's surface can be simultaneously quantified or measured in respect of many different attributes. The vertical arrow indicates how an identical pixel in the grid

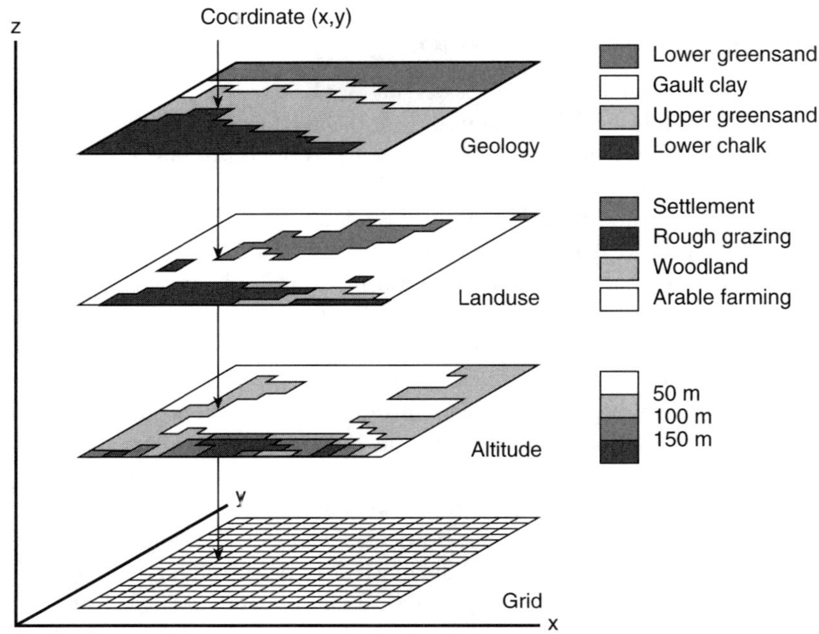

Figure 5.3 Representation of geographical features in layers using the raster model.

forms part of the space occupied by a different feature in each layer. The computer cannot present nor the human eye visualize or interpret all these, and more, attributes at the same time. Thus, multi-attribute displays are achieved by merging information from different layers into new assemblages of unique combinations of values. For example, a possible combination in this case would be cells with altitude 50–100 m, arable land and Upper Greensand.

The resolution of the raster grid (the size of the squares) controls the precision with which the extent and location of a feature can be defined (Wehde, 1982). One advantage of square grid units over the alternative hexagonal or triangular ones is the ease with which the resolution can be altered. Smaller units will normally provide greater accuracy, at the expense of generating a larger dataset and data redundancy, which may be particularly undesirable when the spatial extent of the features is large. Compact methods of storing rasterized data (chain, block and run length codes, or quadtrees) have been devised to reduce data redundancy (Mark, 1986; Waugh, 1986; Samet, 1990; Stuart, 1990). Figure 5.4 illustrates with respect to the Upper Greensand rock types in the geology layer of Figure 5.3 how the grid cells that comprise this feature can be coded according to these data compaction methods. The feature consists of 94 grid cells at one extreme, maximizing data redundancy, the row, column and attribute could be stored (282 data items altogether), the chain code system would reduce this to 37 and structured hierarchy of the quadtree approach would require 34 items to be stored. The raster model has become increasingly popular for storing and displaying geographical data, although the principles of the system underly some of the earliest attempts at computer cartography in the 1960s with the SYMAP and CAMAP software. It is particularly suited to internet-based mapping systems where the pixel locations and their code values can easily be transferred.

The vector model uses basic spatial or geometrical elements (points, lines and areas) to define or construct geographical features, and the system is therefore closer to the way in which traditional maps represent information. Figure 5.5 provides a vector map representation of the same three classes of geographical phenomena shown in Figure 5.3. As with the raster model, these point, line and area elements are positioned by means of pairs of co-ordinates in respect of the Earth's surface by reference to a regular square grid. The vertical arrow indicates how an identical set of co-ordinates forms part of the boundary line defining the space occupied by a different feature in each layer. In contrast to the raster model, the vector approach generates relatively small data files, since co-ordinates only need to be stored for locations where a point, line or area defining a feature occurs on the Earth's surface. In other words, it is not necessary to store the co-ordinates for all of the territory enclosed by a set of boundary lines, nor for straight segments between points on a line. It is simply necessary to store and/or display those which are required according to the desired level of conceptual and graphic generalization (see below). However, when vector data are used within a GIS, data manipulation can be performed which requires the software to store the 'missing' co-ordinates implicitly within the geographical database so that the intersection of two or more different features (e.g. intersection of two different types of polygon) can be achieved. There are a number of ways in which the vector model can be translated into a particular data structure, although there is a basic division between spaghetti (unstructured) and topological methods. The so-called spaghetti structure treats each digitally encoded line as a separate object in space, existing without reference or connection to any other line. The data are thus said to be unstructured because the various individual lines which may be required to construct geographical features, such as a set of administrative boundaries or a road network, are not associated with each other and recognized as objects in the digital data. The system knows how to reproduce the position of the lines in respect of its own location grid on a screen or plotter, but it does not allow individual lines to be identified or related to each other. These relationships are manually added to the data, although some scanning systems partially automate the process.

Geographical data in such a system are said to lack topology, which means that no definition of the spatial relationships between geographical features and their constituent point, line and area elements are stored in the metadata. The Dual Independent Map Encoding (DIME) system devised by the US Bureau of the Census, and its successor (TIGER), were two of the more advanced vector data structures to be developed. These systems, which form the basis of the data structure in many contemporary vector GIS software packages (e.g. Arc/Info), incorporate information about adjacency and connectivity by starting from the principle of capturing the begin and end nodes of a line segment and the direction of travel thus generating the

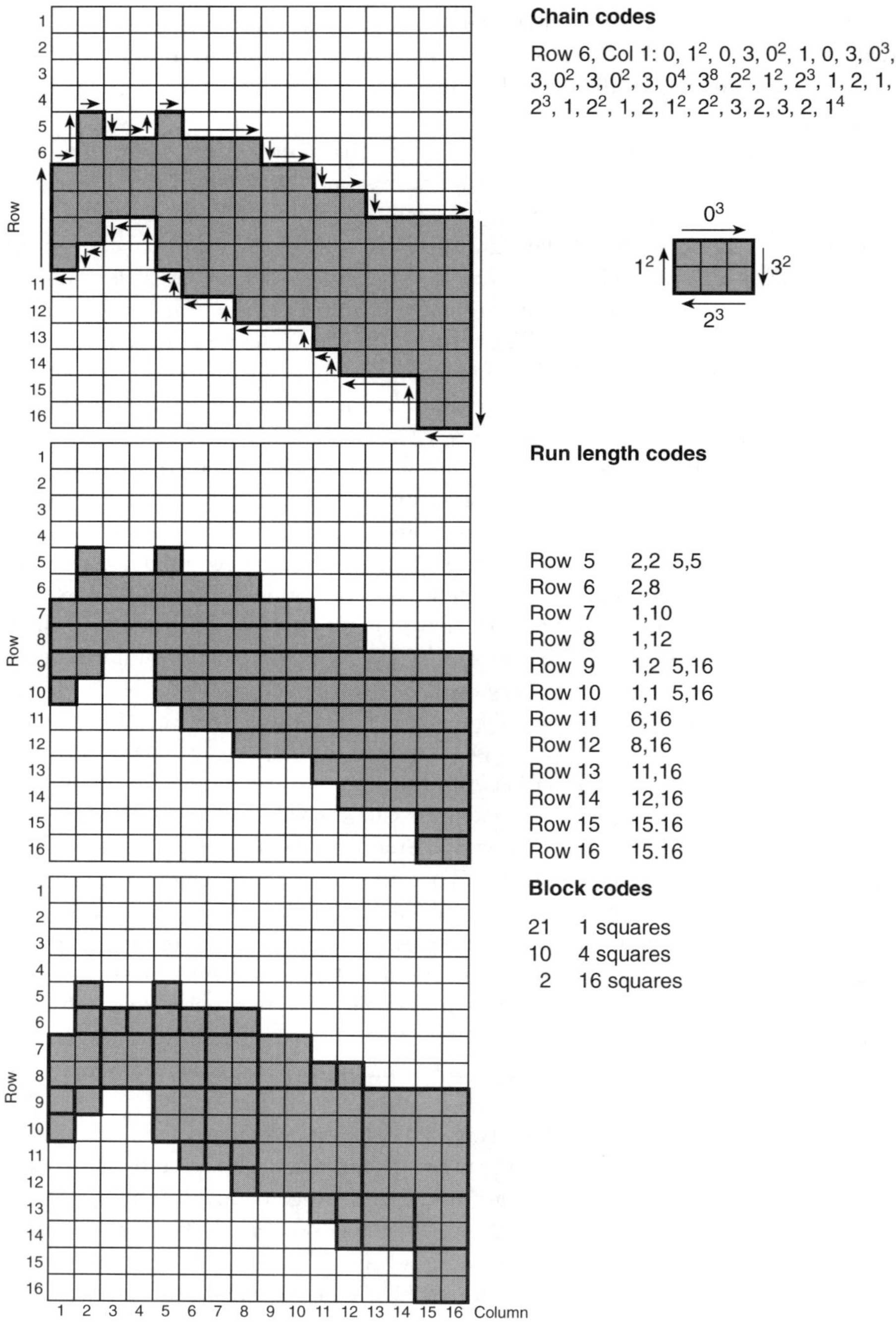

Figure 5.4 Compact methods of data storage using the raster model.

Quadtree codes

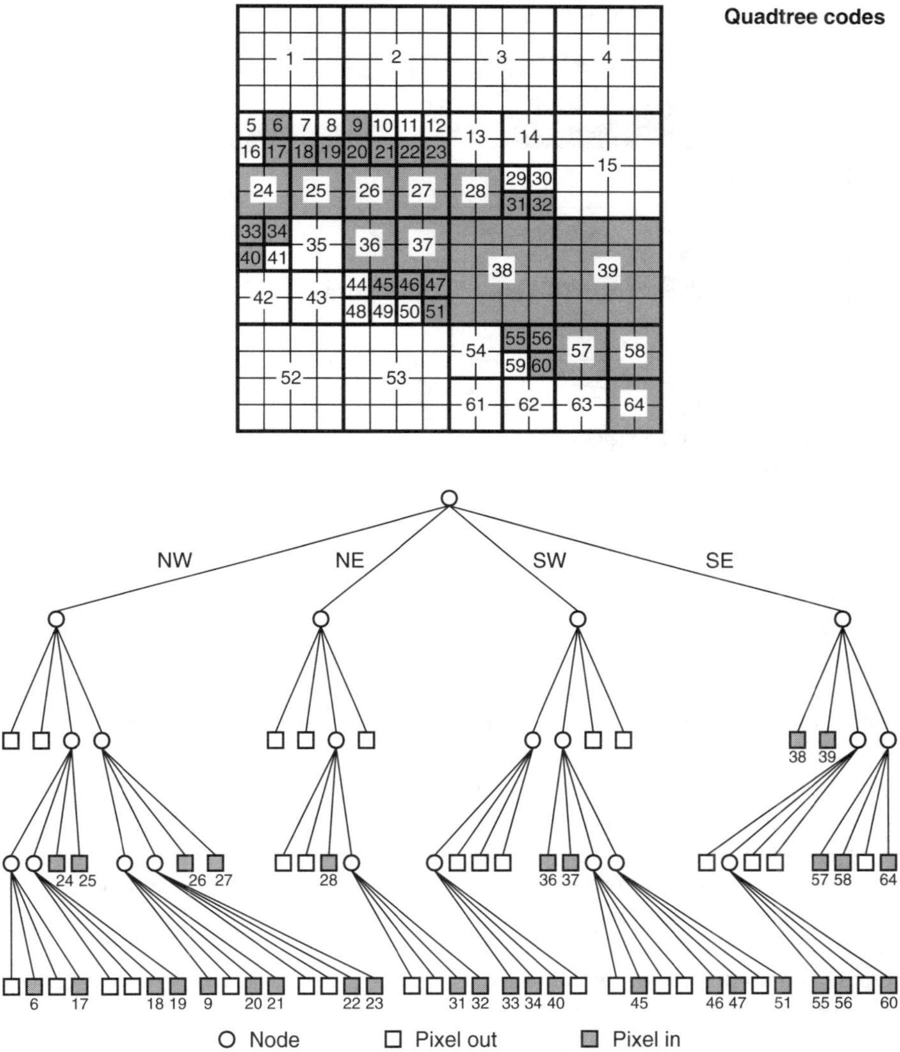

areas occurring on the right and left. This system provides the greatest amount of flexibility in terms of being able to produce different graphic images of the same area. Figure 5.6 illustrates the information required to store the solid geology units in Figure 5.5 using the full topological model. This structure entails enclosing the features within an 'envelope' polygon, in this case corresponding to the entirely arbitrary limits defined by the square map tile. At each of the nodes arrow pointers indicate the connection between an incoming line and the node, and the node and each outgoing line. The polygons themselves, including the 'envelope' polygon, are connected by means of similar pointers between their seed points. The lower part of

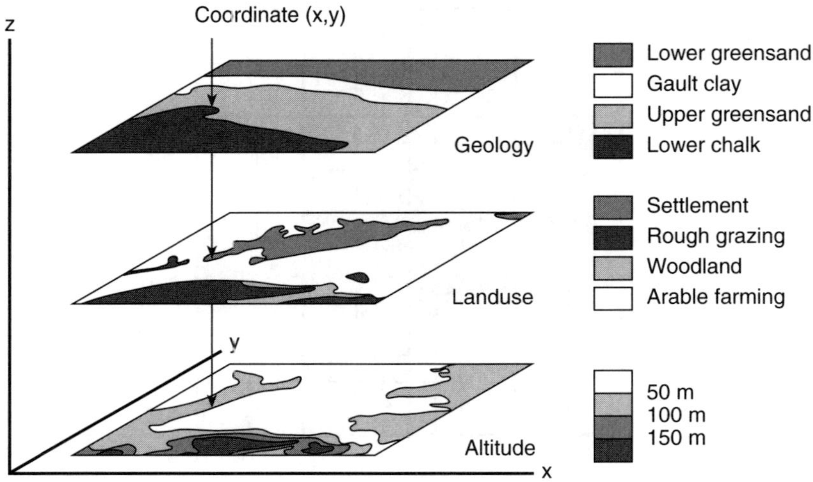

Figure 5.5 Representation of geographical features in layers using the vector model.

Figure 5.6 provides a simplified topologically connected data structure corresponding to the vector map. It illustrates how the complex map structure is broken down into its elemental constituents, namely the co-ordinates for individual vertices, some of which are begin/end nodes, which together make up lines that have angular connections with nodes and form the boundaries of polygons that have a left and right neighbour, which may be the outer envelope.

The second issue related to the capture and conversion of cartographic data concerns the procedures for creating digital representations of mapped information. This is mainly an issue connected with computer hardware, which divides into the digitizing and scanning approaches to the capture of map data. Although scanning and digitizing to some extent parallels the division between the raster and vector models, there are some vector scanning devices as well as raster ones. As with most types of computer hardware, there are qualitative and technical differences between relatively more and less sophisticated devices, which are naturally differentiated in terms of cost. A digitizer is a flat surface table or tablet connected to a serial/parallel port on a computer with an embedded mesh of wires arranged along Cartesian axes. An electrical charge will pass along these wires when the cursor or puck button is pushed, which generates X and Y digitizer co-ordinates, that can be converted into real-world co-ordinates (e.g. latitude and longitude or British National Grid co-ordinates). Manual digitizing also enables the entry of topological information using appropriate alpha-numerically labelled buttons on the cursor, by clicking on or pressing menu buttons on the screen or customized digitizing table. As part of the map preparation procedure the source map(s) will usually have been annotated to indicate such information as which points or lines are to be captured, the direction of travel along a line, the position of label points and the unique (numerical) identifier for each geographical feature. Manual digitizing is a time-consuming and laborious

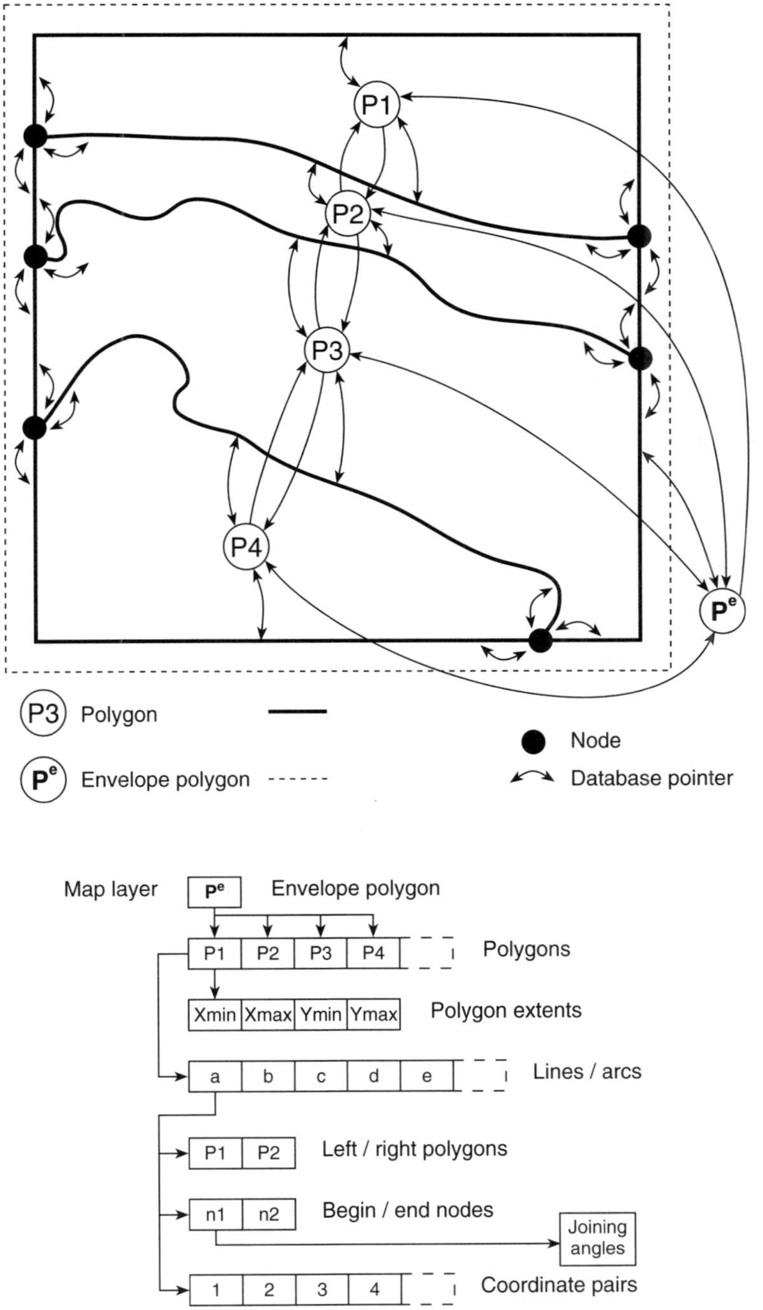

Figure 5.6 Full topological structure for data storage using the vector model.

operation, which can be subject to operator error. The various types of digitizing error include undershoots, overshoots, omissions, spikes and double point or line digitizing (Burrough and McDowell, 1998). The collective effort expended on data capture over the last two decades means that manual digitizing is now a less common activity than formerly, since either much of the existing cartographic data sources have already been encoded or new data capture techniques (e.g. GPS) have enabled digital co-ordinate data of geographical features to be obtained directly.

The precise details of the digitizing operation vary between different digital-cartography and GIS software. Once the requisite good quality sources maps have been obtained, digitizing can proceed, having due regard to issues of copyright examined in Chapter 3. The digitizing procedures to be followed need to be established, for example the sequence in which map features will be captured, the naming conventions to be used and whether points or line segments will be collected first. If the project entails digitizing multiple map sheets, then it is important to decide how these will be progressed. It is useful to mark up the source map(s) to clarify any line intersections as nodes, to specify nodes on island polygon boundary lines and to label and locate polygon seed points.

The principle of capturing data by means of a scanner involves the passing of an optical sensor over the map to register differential light intensity (or colour value), which is converted into digital values. High-resolution scanners are freestanding devices with a rotating drum to which a full size printed source map is attached. The sensor moves in the perpendicular plane to the rotating drum, thus capturing digital data from the map in a series of narrow rows containing small squares. Lower resolution, small format (A3 or A4 size) desktop scanners register light intensity for discrete data points line by line over the map and thus create a rasterized data file. The microfilm/fiche scanner, such as the Digital FilmScan™ DMS4000 supplied by Infographix, enables microfilm or microfiche photographic images of graphics and text to be scanned as raster images directly into computer files. Maps scanned in this way can be brought into GIS software for further processing. Although scanning avoids many of the potential errors of digitizing, one of its main drawbacks is the need to post-process the scanned data file in order to make sense of the scanned lines. Scanning creates rasterized digital images of the source material that may need to be converted into a vector format and to have topological information added.

5.1.3 Map Scale

According to the definitions given above maps can be thought of as abstractions of reality in the sense that only a particular selection of geographical features will be shown on a given map. They are also abstractions with regard to the amount of detail and their geographical extent, since a map cannot exactly reproduce at the same size the reality of the area of the Earth's surface which it covers. The important concept here is map scale, which is the ratio between a distance on a map and the associated distance on the Earth. The scale of a single map produced by means of printing, displayed on a VDU screen or drawn by a plotter, will be constant across the entire area. Should parts of or complete maps be photocopied, then the optical system in the copier will introduce distortions. This type of error means that photocopied maps should be digitized or scanned with caution, irrespective of issues related to breaches of copyright. Maps are published at a variety of different scales, which are denoted by a representative fraction (1 : 2500), by a graphic figure (an accurate scale bar) or, more rarely, by textual description (1 cm to 2500 cm (25 m)). The scaling of maps can be thought of as carried out on a continuum, where for example a given distance (e.g. 1 cm) on a map drawn at 1: 1250 denotes shorter real-world length than the same distance on a map at 1: 125 000. In this case 1 cm on the first map would represent 12.5 m whereas on the second it would denote 1250 m. However, it is more common to think of maps as being either small or large scale, although these are clearly relative terms. There is no universally agreed cut-off between small- and large-scale mapping, and in some cases the term intermediate mapping has been introduced. Figure 5.7 shows typical examples of raster topographic mapping at large (1: 50 000) and medium/small (1: 250 000) scales, which have been reproduced from the Canadian mapping agency's Toporama system (see Chapter 8). The descriptive terms large and small sometimes cause confusion when used with reference to maps: a small-scale map shows a small amount of detail for an area of large geographical extent; a large-scale map contains a large amount of detail for an area of small extent.

The issue of map scale is important from the GIS analyst's point of view because the scale at which the cartographic data have been collected and compiled into a map affects their usefulness for analysis and interrogation, as well as for display purposes. This issue connects with conceptual and graphic generalization, since as a general rule it is both feasible and permissible to remove detail selectively from a map that has been produced at a large scale in order to generate one at a smaller scale, whereas the reverse is not advisable and often impossible. These changes of scale may affect both the topographic and thematic content of a map and there is clearly the potential for discontinuities to arise when the scale of the data during the original collection events were different for these components. Suppose that the delineation of Population Census enumeration districts was carried out on base topographic maps at a 1: 10 000 scale, as the map scale is made progressively smaller the physical extent of these units in absolute terms on the printed or digital map becomes smaller until a point is reached where the spatial units and the attribute information they contain are indistinguishable (Figure 5.8a). The problem can be overcome by removing detail from both sets of data. However, because the nature of the underlying scale of the two types of data is different, the cartographic data are on a continuous scale, whereas the attribute data are on a nominal scale and can only be aggregated to a larger spatial unit in the hierarchy (e.g. enumeration district to ward), less than optimal results may be produced (Figure 5.8b). If the lowest georeferencing of the attribute data was on the same numerical scale to the digital topographic data (i.e. dwellings referenced at 1: 10 000), then it would not be necessary to change between spatial units in such large steps as between an enumeration district and an electoral ward. It would be possible to construct a continuous surface from the georeferenced point locations representing the attribute data. Figure 5.9 shows such a surface for part of the area

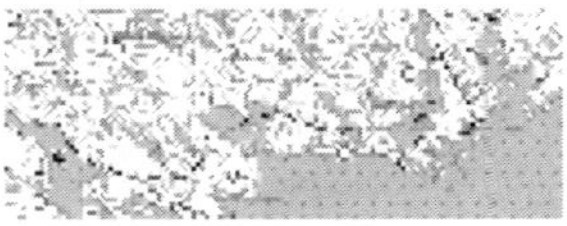

Scale 1:250 000

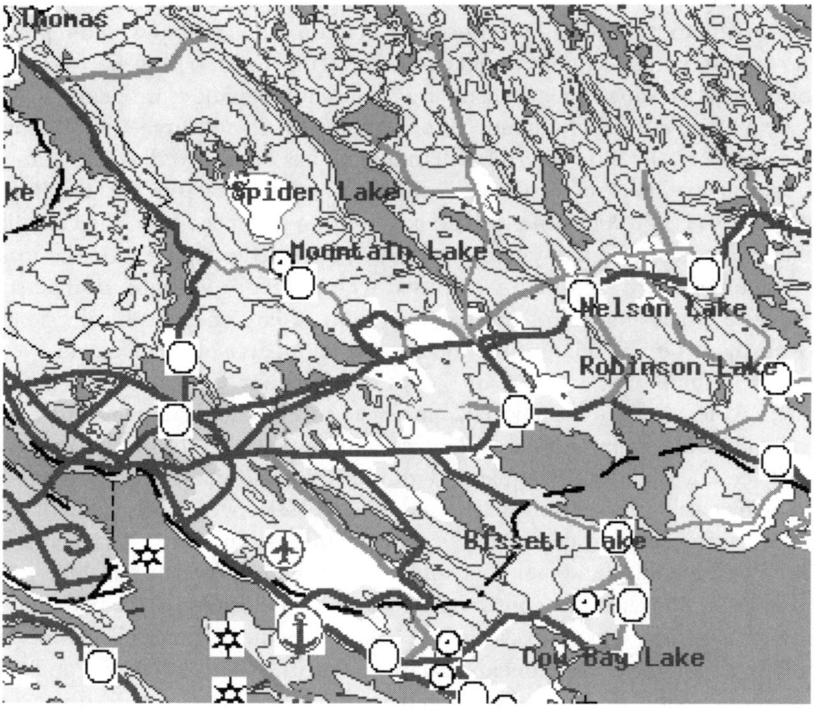

Figure 5.7 Examples of small-scale (1 : 250 000) and large-scale (1 : 50 000) topographic maps for Canada. © Her Majesty the Queen in Right of Canada 2001.

covered in Figure 5.8. The trend surface gives an impression of highs and lows of the mapped attribute, but details have become blurred.

5.1.4 Data Generalization – Graphic and Conceptual

The process of generalization with respect to cartographic data reduces the amount of detail contained in a map at a given scale. The term is normally used in connection with the design and compilation of maps. The process of generalization is to a large extent concerned with visualization of geographical data and map design rather than analysis, and therefore a comprehensive account of the issues is left to others (Dent,

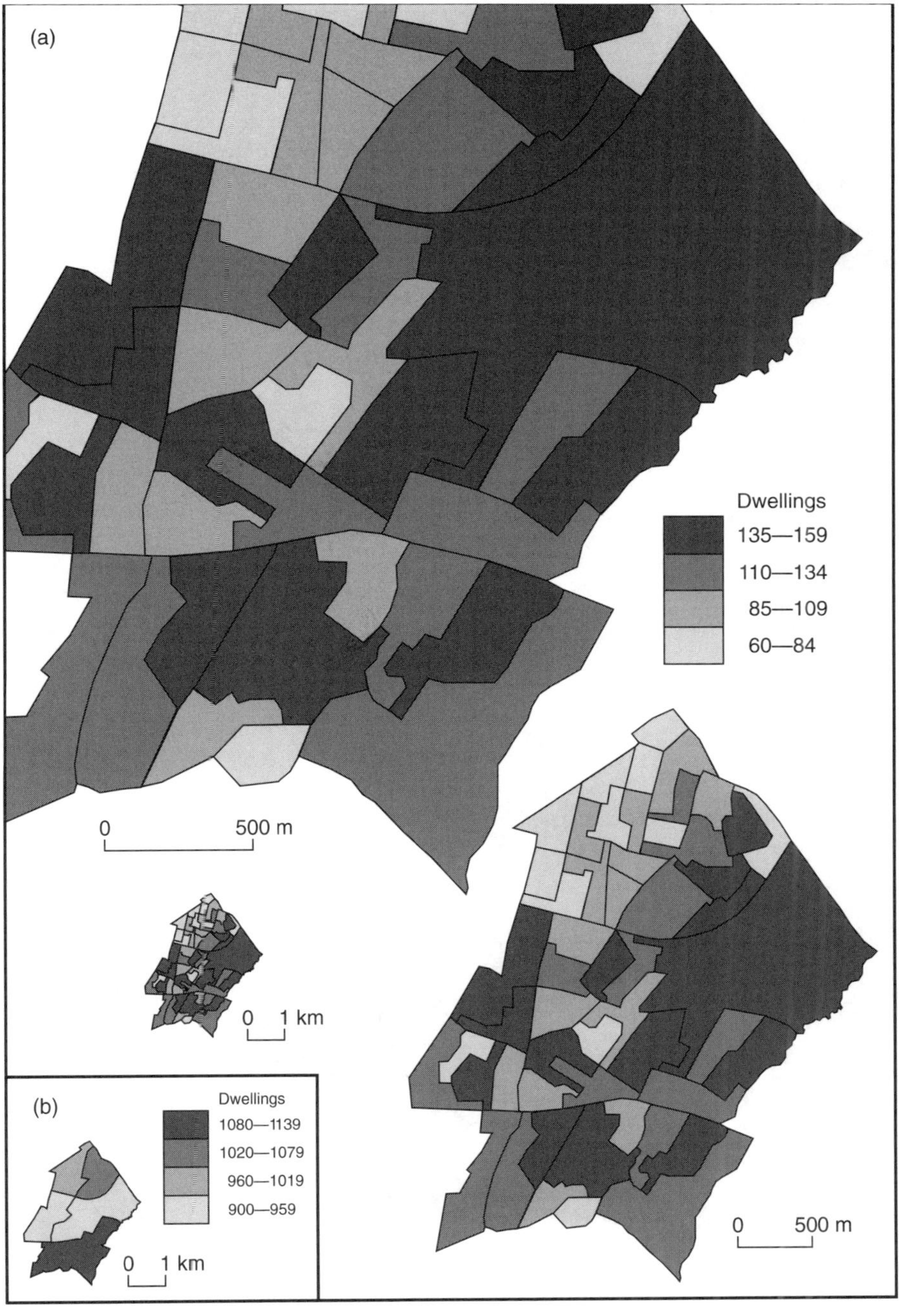

Figure 5.8 Effect of scale changes on the display of attributes for geographical features. (a) Same spatial and thematic information shown at three different scales, (b) adjustment of spatial unit to compensate for scale change.

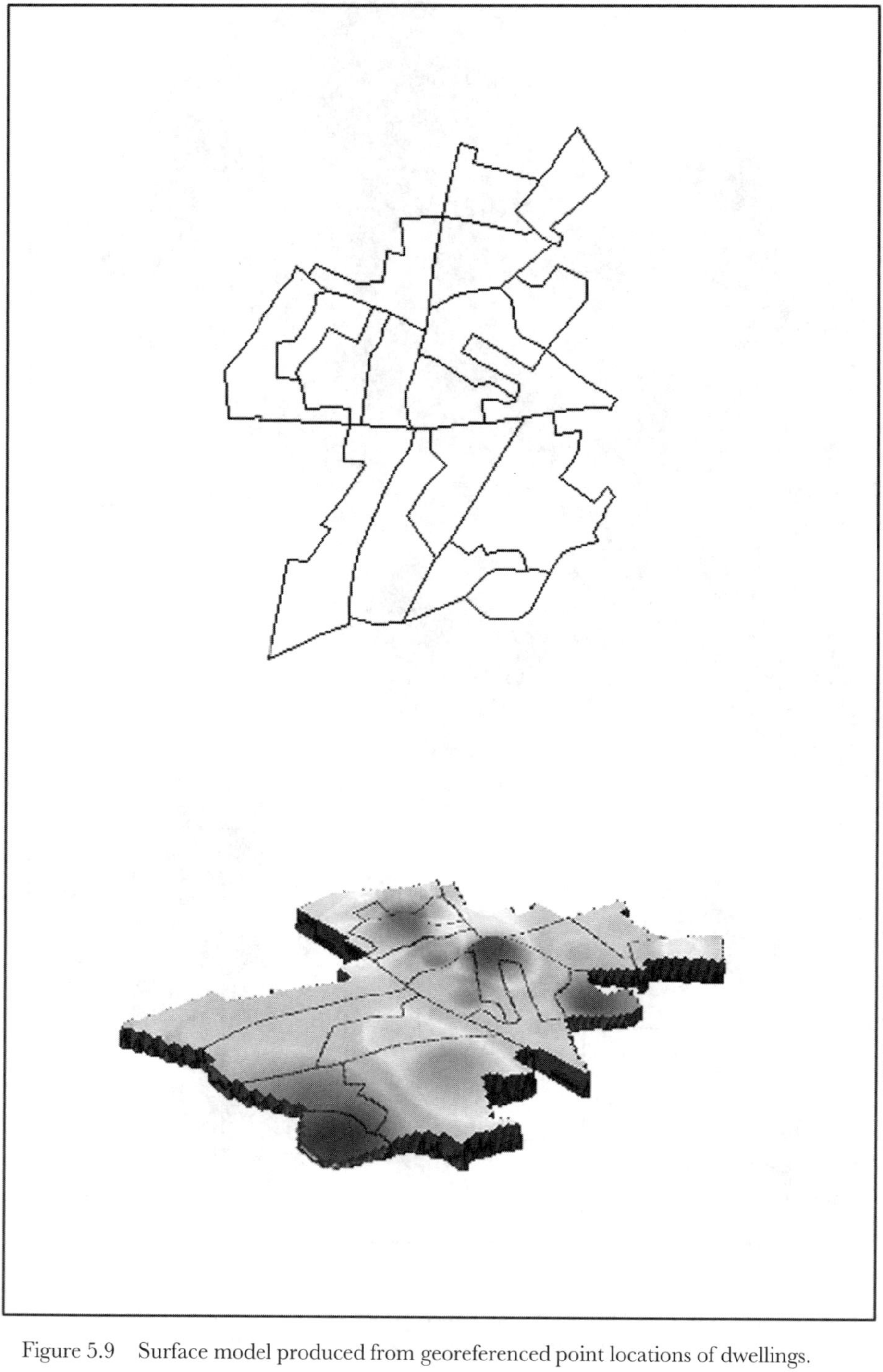

Figure 5.9 Surface model produced from georeferenced point locations of dwellings.

1993; Kraak and Ormeling, 1996). However generalization is pertinent to users of maps as sources of geographical data in certain circumstances, since some of the techniques can be applied when data are captured from existing paper maps or extracted from digital files. Furthermore, the readily available facility of zooming in and out of digital map displays creates a false sense of authority and power which shows little regard to the validity of the exercise in the hands of the untutored user.

Two types of generalization, conceptual and graphic, are usually recognized (Kraak and Ormeling, 1996) and can be achieved by a number of different processes. Selection, simplification, enlargement, displacement and merging all relate to graphic generalization, whereas the conceptual category includes merging, selection, symbolization and exaggeration (Figure 5.10). In the context of accessing existing cartographic data, whether in analogue or digital form, for GIS analysis the main considerations relating to generalization concern the simplification of detailed information taking into account the purpose for which the captured data will be used. GIS analysis often involves integrating two or more data sources; clearly if the results of this process demand a high level of precision and accuracy then the scope for generalization may be limited. On the other hand, if the captured data are intended for a more supportive, pictorial background function, the scope for generalization may be greater.

The manual digitizing operation (see Section 5.1.2) provides the user with the opportunity to determine by eye which points along a line, and therefore the amount of detail, will be captured. Since this will inevitably involve subjective judgement on the part of the operator, an alternative more sensible strategy is to capture all detail in the source data as fully and accurately as possible and then apply automatic generalization algorithms from within the GIS software. Such automated simplification can also be used on scanned data, where the amount of detail captured is defined by the size of the raster pixel, which can be as small as 0.1 μm. When obtaining digital cartographic data for use in a GIS, the data supplier may offer different generalization options to customers, which will affect not only the appearance of the displayed map and the data's analytical potential, but also product price. Another generalization technique which may also be appropriate when acquiring cartographic data from maps or using existing digital files is displacement. This may be necessary when attempting to edge-match different map tiles and features such as roads, rivers, administrative boundaries and other linear or polygonal features do not align perfectly. The displacement of one or both lines so that they line up across the map join in such cases may either be undertaken at the digitizing stage or subsequently by means of snapping mismatched lines together within computer-cartographic or GIS software.

5.1.5 Storage Medium

The medium of storage of the earliest known maps was a cave wall; whilst it is unlikely that these will be directly captured for use in GIS analysis, photographs may well be scanned for computer-based investigation. During the centuries since these early

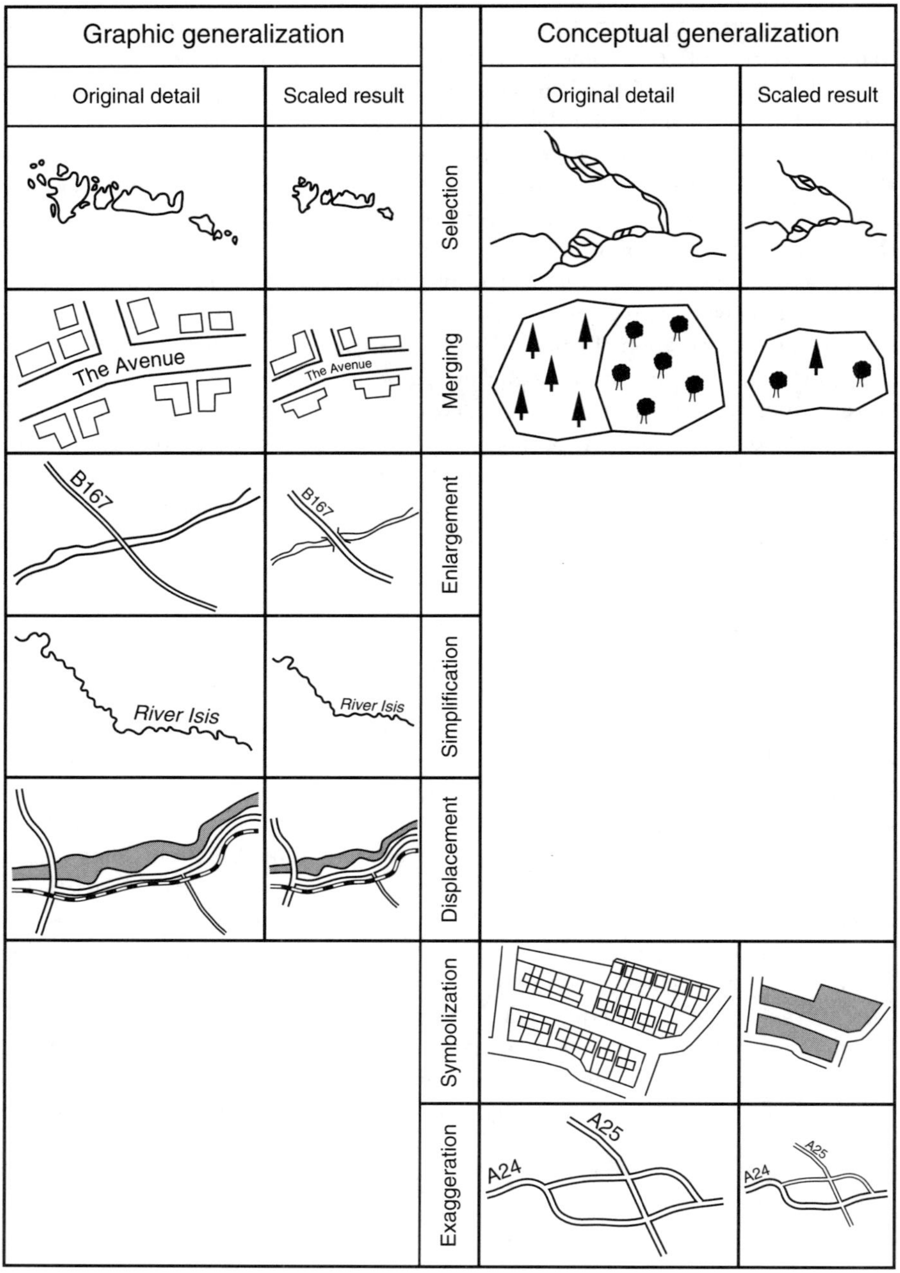

Figure 5.10 Graphic and conceptual generalization of map information.

maps were made, the most common storage medium has been high quality paper. Many of the maps which have been digitized or scanned during the last three decades of digital-cartography and GIS analysis have been sourced from such printed documents. From an archival or storage perspective paper maps require considerable storage space and may be prone to damage and discoloration. From the GIS analysis point of view, paper maps constitute an important, if not sometimes the only, accurate source of geographical data about the Earth's surface, particularly if an historical perspective is required. As more and more mapping organizations undertake their work using information technology, so the storage medium for cartographic data sources is turning increasingly towards electronic media, including disks, tapes, CD-ROMS, digital archives and the internet.

The provenance of any secondary source of geographical data is important for GIS analysis. In the case of traditional printed maps the necessary metadata information relating to the direction of magnetic north, the map projection, the date of survey and data collection, the map scale, the symbolism, ownership and other essential details were included as part of the map's design and layout. In the case of cartographic data stored in a digital format for distribution to users this information may not be directly visible or accessible from the displayed digital map, but may be held in a separate metadata file (see Chapter 3). It is clearly important for users of map data from secondary sources to be conscious of the need to assess and understand any limitations or restrictions relating to a digital map.

5.2 TOPOGRAPHIC MAPPING

Most people first encounter maps in the form of a school atlas in which there is a traditional division between physical or political maps (see, for example, Hammond Incorporated, 1996). The former represent the physical features of the Earth's surface, including coasts, rivers, upland areas, lakes, seas and oceans, and the location of principal human settlements. They show where these features are in respect of their precise geometrical relationships with each other, but apart from cartographic text such as a name and, in the case of land and water bodies, perhaps some indication of differential altitude or depth respectively, they show nothing about the attribute characteristics of the features. This is the task of political maps, so-called because they show the world divided up into the arbitrary political land parcels that constitute nation states. It is these political units that have been traditionally viewed as providing the basis of spatial variation in human phenomena and hence form the entities whereby we might examine patterns of population, economic activity and social conditions. The terms physical and political map have been superseded by topographic and thematic. However, even this terminology is no longer appropriate or entirely necessary in the brave new world of computer mapping and GIS, since 'in a digital environment the differentiation between topographic and thematic maps is less relevant' (Kraak and Ormeling, 1996: 44). The division between physical and political maps is less important than formerly because we no longer rely on printed versions of maps to provide the definitive statement of where features on the Earth's surface occur or to describe the characteristics of these features. The time scale for producing such definitive statements and descriptions has been compressed and we now have the facility to represent changes as they occur. GIS technology provides the user with greater flexibility for integrating and overlaying digital representations of geographical features and thereby to create new entity classes based on the myriad of potential combinations that could be obtained from the 'raw' input features. Nevertheless, from the perspective of examining cartographic sources of geographical data for use in GIS, it is still relevant to distinguish between topographic and thematic mapping.

In most countries the responsibility for maintaining the national topographic database and for keeping the national map series up to date lies with a government supported agency. In the case of Great Britain this task falls to the Ordnance Survey (OS) and in Northern Ireland to the separate Ordnance Survey for Northern Ireland (OSNI). Until 1990 these were directly under the control of central government, but since that year they have enjoyed executive status, giving them greater autonomy over their activities and policies. The equivalent body in the USA is the United States Geological Survey and in France it is the *Institut Géografique National* (IGN). The development of topographic mapping and land survey in many countries is linked with military campaigns that occurred long before such formal mapping agencies were established. As archived documents, these ancient maps are of interest to scholars of military history and historical cartography, but are less relevant to the purposes of this book.

Some contemporary mapping agencies also have their origins in such military mapping. The OS was founded in 1791 in response to the threatened invasion of southern England from France. The maps were produced for counties in this region at a scale of one inch to one mile and proved so successful that plans were developed to extend the coverage to the whole of Great Britain. In 1825 a major survey commenced in Ireland at the new one inch to six miles scale (1: 10 560) for taxation purposes. During the middle decades of the 19th century mapping was carried out at other scales according to the general characteristics of the land area, so that in built up areas where the population exceeded 4000 a scale of 10 feet to the mile was employed. Similar disputes over the suitability of mapping scale still raged in the 1930s when the three basic map scales were established by the Davidson Committee 1935–38, which also recommended that the whole of Great Britain should be retriangulated and surveyed on a metric National Grid. The operations of the OS were dominated by implementing these and other recommendations over the next 40 years, only delayed by World War II. However, maps based on the new National Grid continued to be published using the familiar non-metric scales, for example 1: 10 560 until the 1970s.

In recognition of the likely future direction of cartography and the emerging role of computer technology in this field, the OS embarked upon a programme in 1973 to convert the National Topographic Database into a digital format. By the mid-1980s, some user organizations, dissatisfied with the rate of progress and with what was perceived as excessive detail in the feature coding employed by the OS in its National Transfer Format (NTF), had initiated their own digitization of existing topographic maps. This problem was identified by the Chorley Commission (HMSO, 1987) and it recommended that the OS accelerate its digitization programme. This recommendation, together with pressure from users and the OS's acquisition of executive status, had the desired effect and by 1995 all basic scales of OS topographic mapping had been digitized.

The British National Topographic Database is maintained by the OS and comprises more than 280 000 maps at seven main scales, which are estimated to contain over 200 million landscape features. Maps from this database have been published at either the basic scales (1 : 1250; 1 : 2500 and 1 : 10 000) or the derived scales (1 : 25 000; 1 : 50 000; 1 : 250 000 and 1 : 625 000). Basic scales maps are the largest available for any area and account for about 97% of the map sheets. Maps in the second group are published at a scale that is derived from the basic scale mapping. In other words, since the map sheets cover a larger area, they can be conceived as 'cut down' or generalized versions of the basic scales mapping in which certain features have been suppressed and line drawing has been altered to allow for the change of scale. Table 5.1 illustrates the differences in the number of feature codes used in three of the OS digital data products. Road line features illustrate how scale affects the nature of digital information stored about geographical features changes with scale. The 1 : 250 000 data has 44 codes to identify different categories of road (e.g. motorway, A roads and minor roads), their status (e.g. under construction or planned in the case of motorways) and whether they are within tunnels, on overpasses or on the ground surface. The large-scale digital data includes only two road feature codes, but these relate to the metalled edges and centre lines.

Table 5.1 Comparison of number of feature codes used contrasting scale Ordnance Survey digital map products

Product	Land-Line Plus®	Meridian2™	Strategi®
Scale		1:50 000	1:250 000
Urban areas	1:1250		
Rural areas	1:2500		
Remote areas	1:10 000		
Total features	50	50	214
Road features	2	4	44
River line features	1	3	6
Woodland area features line boundary	1	1	1
Railway line features	2	2	5
Administrative/electoral area boundary and coastline line features	5	4	5

Note: Apart from Total features, the counts exclude NTF text codes, seed points and neat lines between map tiles.

In recent years the OS has become proactive in developing a range of products from its basic topographic mapping. Table 5.2 summarizes the main OS products currently available. One important group of products has been marketed as Business Geographics since 1999 and is designed to fulfil mapping requirements ranging from obtaining a small-scale (1 : 1 000 000) map as a backdrop or for magazine and brochure illustrations through mid-scale (1 : 50 000) raster maps for land use planning, through to more specialist products for applications such as route planning and geodemographics. There is also a group of Detailed Products of which the most important is Land-Line®. This comprises detailed topographic vector mapping data with national coverage at three scales of survey:

- 1 : 1250 for urban areas (500 m × 500 m)
- 1 : 2500 for rural areas (1 km × 1 km)
- 1 : 10 000 for moorland areas (5 km × 5 km)

There are 228 876 tiles edgematched to facilitate the creation of seamless displays across multi-tile regions. Two versions of Land-Line® exist, one containing only features of the built environment (e.g. houses, factories, roads and railways) and Land-Line Plus® that also includes landscape features of the natural environment. The position of the Ordnance Survey within the structure of digital geographical data providers within the UK is examined in Chapter 8, where comparisons are made with a selection of other countries.

Northern Ireland has its own mapping agency, the Ordnance Survey of Northern Ireland (OSNI), and since 1955 all its maps are based on the Transverse Mercator

Table 5.2 Summary of Ordnance Survey digital data products, 2001

Land-Line®/Land-Line Plus® national coverage at 1 : 1250, 1 : 2500 or 1 : 10 000 depending on area ongoing update
Vector data – Meridian2™ national coverage at 1 : 50 000
Vector data – Strategi® national coverage at 1 : 250 000 updated annually
Vector data – Land-Form PANORAMA® national coverage at 1 : 50 000 contours and digital terrain models (10 m interval contours and DTM grid at 50 m horizontal intervals
Land-Form PROFILE™ national coverage 1 : 10 000 contours or digital terrain model
Place name gazetteer – all names appearing on 1 : 50 000 Landranger map series
Vector data – BaseDate.GB® national coverage at 1 : 625 000 updated annually
ADDRESS-POINT™ – national coverage georeferenced postal address data
Code-Point™ grouped ADDRESS-POINT™ co-ordinates by postcode
OSCAR® road centre line vector data for motorable roads
Boundary-Line™ national coverage administrative and voting boundaries
EDLINE 1991 Census of Population enumeration district and output area (Scotland) boundaries
SABE – Seamless Administrative Boundaries of Europe cover of 26 European nations to NUTS 5 level
Raster data national coverage 1 : 50 000 ongoing update taken from Landranger maps
Raster data national coverage 1 : 10 000 ongoing update colour or monochrome
Raster data historical mapping from previous printed maps series in conjunction with Landmark Group

projection covering the whole of Ireland, known as the Irish Grid, with a true origin at Latitude 53° 30′ N and Longitude 8° 00′ W. The island as a whole is covered by 20 100 km squares, each of which is labelled by a letter. The national mapping agency in the USA, the United States Geological Survey (USGS), dates back to 1879, and has accumulated a large collection Earth and biological data in addition to its main mapping activity. Mapping of the USA at the scale 1 : 24 000 (the 7.5 minute series) to produce topographic maps started after World War II and was completed by the 1980s with over 57 000 sheets printed to cover the 48 coterminous states. This fundamental series has now been extended to smaller scales and to cover other US possessions and territories. Historically topographic mapping has also been linked to economic development. The former colonial powers undertook topographic mapping of their dominions as part of the overall process of exploration, domination and administration during the 19th and early 20th centuries. Thus, much of the basic topographic mapping in developing countries was originally carried out by countries such as Belgium, France, The Netherlands, Spain and the UK. Contemporary topographic maps for many of these countries are still published by mapping agencies in their former colonial masters.

5.3 THEMATIC MAPPING

Traditionally the principal difference between topographic and thematic maps is that the former show the location of a range of physical features on the Earth's surface, whereas the latter present the geographical distribution of a single attribute, or occasionally a small number of attributes. The physical features depicted on topographic maps can be thought of as the hooks on which thematic (and graphic) attributes are hung. The topographic information will often form the background for thematic mapping with physical features usually shown in a simplified or stylized form to provide a locational context for variation in the attribute. The primary purpose of a thematic map is not to show the features that occur in an area and their spatial relationships, but to portray the incidence of a particular characteristic in relation to those spatial features. In some cases these characteristics will be connected with the topographic features themselves, for example pollution levels in rivers and lakes, whereas in others they will be attached to derived spatial units, for example population density linked with administrative areas. Dorling's work mapping statistics from the 1991 UK Census of Population (Dorling, 1993, 1995a, 1995b) has explored innovative, cartogrammetric ways of representing the economic and social characteristics of the population in which the underlying spatial units to which the statistics relate vary in size in terms of physical area and population.

Attribute data may be associated with the basic mapping units or primitives, points, lines and polygons that make up geographical entities and also with the composite features defined by combinations of these units. Figure 5.11a presents a generalized topographic map and shows the presence of various point, line and polygon features. These form the primitive geographical units that make up the

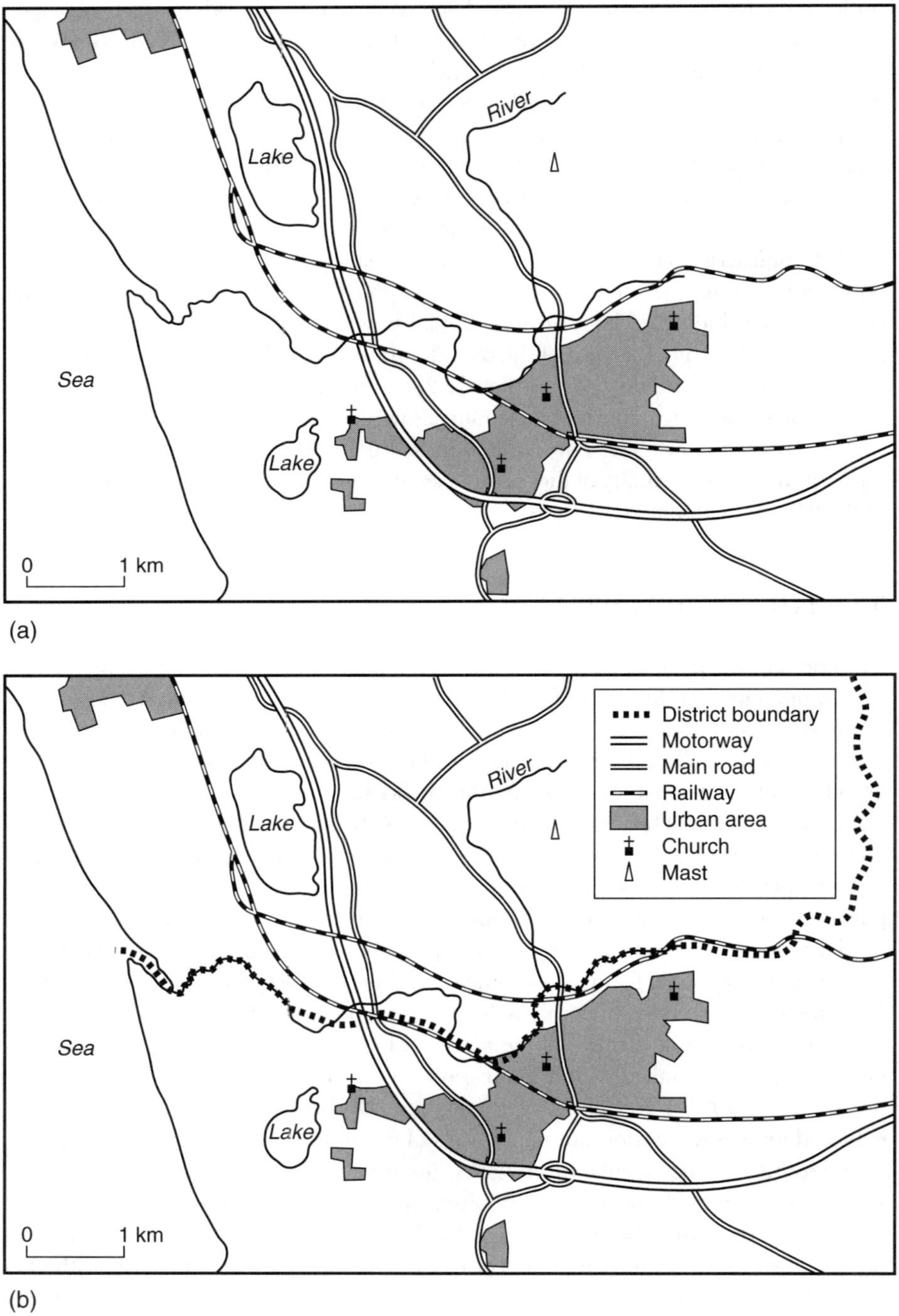

Figure 5.11 Use of spatial primitives to form primary and secondary topographic features. (a) Simplified topographic map, (b) use of river line to form part of administrative boundary.

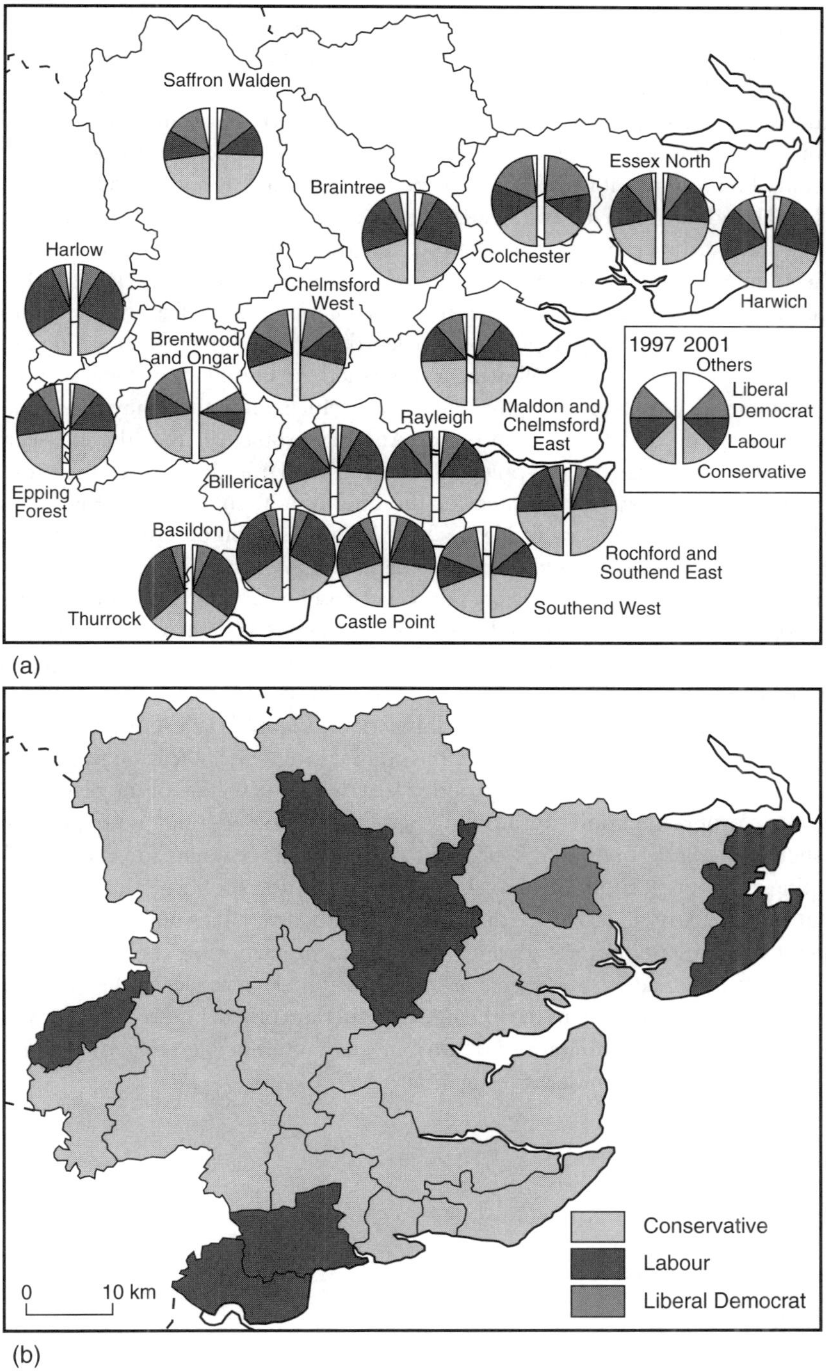

Figure 5.12 Problems of deriving numerical data from thematic maps. (a) Party share of the vote in 1997 and 2001 UK General Elections, (b) the political map of Essex constituencies resulting from the 2001 UK General Election.

topography of the area and depending upon the purpose and scale of mapping may be selectively included or excluded from the cartographic visualization. However, these topographic features may also be used for part or all of their extent to define the geometry of other geographical features. Figure 5.11b illustrates how the river running east–west across the lower part of the map has a local administrative unit boundary line coincident with it for some of its length from the coast inland and then this line abandons the river and heads northwards. This boundary line is itself also likely to have a dual purpose, since various lengths or sections during its extent will connect different local authority wards, which form a lower tier in the British hierarchy of governmental units. In some cases the topographic features constituting parts of the boundaries of such polygons are less distinct than rivers or coastlines, such as watersheds or lakes. When following a watershed the boundary line will seemingly traverse open space of a topographic map, but careful examination may reveal that it connects the summits of mountain peaks. Another problem is that the connection between the boundaries of the polygons and the topographic features may not be permanent, which may be indicated for the intermittent connection between river and local authority boundary in Figure 5.11b. This question of temporal change in geographical features is examined in more detail in Chapter 9.

Thematic maps in themselves may not be very useful as sources of data for further geographical analysis. They should really be considered as devices to communicate the outcome or end product of such analysis. Figure 5.12 shows two types of thematic map that depict information about voting patterns in parliamentary constituencies in Essex in the 1997 and 2001 UK General Elections. Figure 5.12a shows by means of split pie charts the percentage share of the votes cast on each occasion for the three main and all other political parties grouped together. The sectors of the pies enabled the relative proportions and overall changes to be determined, but it is not possible to use such a thematic map as a source for the exact percentages for each political grouping in either election. Figure 5.12b provides the new 'political' map of the area by showing the party that won in each of the constituencies. This portrays the pattern of political representation in terms of which political party won each constituency, but it does not indicate the closeness of the result. These examples of thematic maps not only show how some maps may be more useful than others for obtaining raw data counts, but also that combinations of attributes are often needed to provide a fuller account of the attribute under investigation.

Chapter 6

Censuses and Surveys

6.1 CHARACTERISTICS OF CENSUS AND SURVEY DATA

In the early stages of GIS development there was a tendency to associate the technology with geographical data that can be broadly regarded as environmental or 'physical' in their nature. Indeed the first textbook on the subject *The Principles of Geographic Information Systems* (Burrough, 1986) has the infrequently recalled subtitle *for Land Resources Assessment.* The effect of this emphasis in early GIS applications was for data derived from censuses and surveys to be of comparatively modest importance in the development of GIS. Recently there has been an almost complete turnaround as the importance of geographical data relating to demographic, economic and social characteristics have become more prominent in GIS. This trend is evidenced by the growing interest in business management, geodemographics and other socio-economic applications (Martin, 1991; Goss, 1995; Longley and Clarke, 1995; Sleight, 1995: CACI Information Services, 1996; Martin, 1996). The most frequently used and well-documented secondary source of data for these types of application is probably the Population Census. However, familiarity with this source may be misleading for several reasons. The Population Census is neither the only census nor the only source of generally 'human' geographical data which can be used in GIS. These data fall into two broad groups: first, those censuses and surveys which are specifically collected for the purpose of answering research or investigative questions; and, secondly, those which are a by-product of an administrative process or a statutory requirement. This chapter examines the characteristics of the first group, while the second are addressed in Chapter 7.

Censuses and surveys are of course not solely concerned with 'human' geographical data, since the terms refer to a statistical point concerning the number of observations and the procedure whereby the units are selected for inclusion in a dataset. The word census is derived from the Latin meaning 100, as in percentage, and denotes complete (100%) enumeration or measurement of observations. Digital aerial photographs and satellite images as described in Chapter 4 can be viewed as censuses of the pixels for the areas they cover and the topographic maps outlined in Chapter 5 were originally produced as a result of a complete land survey, although the heights for only a sample of points are likely to have been measured for the purpose of constructing contours. A sample survey in contrast signifies that a subset of observations are examined. The concept of sampling may also be applied to digital aerial photographs and satellite images when a sample of pixels are selected for the purpose of training the classification and to land survey when only part of an area (map sheet) is covered.

In statistical terms therefore a census relates to a population of phenomena as opposed to a sample which only includes a subset of observations. The significance of this difference is that a census will contain data or yield information about every member of a given, defined population, albeit perhaps in an aggregate form. A census may thus, in theory at least, be regarded as an accurate count or measurement of all observations, although for reasons such as instrument failure or non-response some may be unintentionally omitted (Kiernan, 1994). In contrast, a sample will only provide definite, supposedly incontrovertible counts or measurements for a specifically selected group of observations from the parent population. Sample data may relate to individual observations rather than aggregate units, but in an anonymous fashion. Thus any conclusions or statements based on a sample may be subject to various distortions and inaccuracies generally grouped under the heading of sampling error. One slightly confusing point is that such error may also be associated with a census, for instance when the errors originate from faults with the measurement instrument, for example, if questions are open to differential interpretation.

The specific reasons for or sources of sampling error are numerous, but all essentially relate to the fact that the information collected falls short of some level of accuracy. The general question of accuracy was considered in Chapter 3 in relation to data quality. Sampling error can occur for three main reasons. The means used to collect the data may be at fault which can occur when questions are poorly worded on a questionnaire or census schedule causing confusion, resistance or inconsistency amongst the respondents. A similar situation occurs when a measuring device, such as an automatic data logger, has a fault which causes unpredictably inaccurate measurements to be recorded. It is very difficult for the analyst, particularly when using the data from a secondary source, to determine the extent to which these types of sampling error are present. Sampling error may also result from the sampling procedure itself (see below) which can introduce undetected bias into the data. For example, where sampling involves choosing people in the street for a face-to-face questionnaire survey, the interviewer may select those people who appear willing to co-operate in the survey and ignore the seemingly 'difficult' potential respondents. Yet these omitted members of the population would have equally valid opinions and factual information to impart had they been included. A similar type of bias may arise when physical features are surveyed, for example, reaches of a river which are only accessible by crossing private land or by risking personal danger. A third type of sampling error occurs when the initial recording or the digital encoding of data is carried out incorrectly. Individual phenomena may have been given erroneous values at the time of collection or data originally assigned correct values may inadvertently have errors introduced during the process of conversion into a machine-readable form.

Samples of observations can be drawn from populations in many different ways and although certain procedures are usually recommended, it frequently proves impossible to adhere to these statistical strictures to the letter. The overall aim when selecting a sample is that it should be of sufficient size and as representative as possible of its parent population in order to ensure the statistical validity of the estimates. Such representativeness can be thought of as existing when a minimum number of observations provides an entirely accurate estimate of the population from which it has been selected in respect of the measured attributes and variables. In practice it will rarely be possible to select a sample that achieves this goal perfectly. Far more common is the situation in which a sample provides a set of figures or statistics in which the analyst can have confidence that the results are reliable to a stated degree (e.g. 95% confident).

The general principles of sampling procedures were outlined in Chapter 2. In the case of samples relating to geographical features it is important that each observation should be independent of all others in the population, such that the way one is categorized has no bearing on the classification of the others. If this condition is violated then some measure of positive or negative spatial autocorrelation will be present. This results in observations which are close together having similar or divergent values simply because of their proximity. This may cause problems when carrying out inferential statistics, but the very presence of such spatial clustering may be of relevance when describing the geographical patterning of the phenomena. The sampling of geographically referenced data is an important issue for GIS applications because a sample, which is statistically valid at a national or regional scale, may be unrepresentative of smaller geographical units such as counties, districts, electoral wards or post(zip)codes. National sample surveys, such as the General Household and Labour Force Surveys carried out in the UK, exemplify this problem. The General Household Survey is an annual omnibus survey containing approximately 13 000 households. The annual survey report (Office for National Statistics, 2000) gives full details of the stratified random sampling procedure, which essentially involves stratification by socio-economic group. The survey contains a wealth of economic and social information about households and their individual members, which is available to researchers in a disaggregated form for anonymized households and individuals. Unfortunately the data are not statistically valid below the national and standard region levels. Potential users of the GHS in GIS will soon become aware of this drawback by the lack of sub-regional spatial references in the datasets. Thus the total sample of 15 874 people aged 16 and over in the 13 250 sampled households in the 1998/99 GHS included 1489 aged 75 years and over who lived alone. Although it will be possible to determine those living in each government office region, their precise residential location cannot be identified. These individuals are distributed across the counties, districts and postcode sectors within these regions, however, they have not been selected in the correct proportions to their occurrence in the populations of these geographical units. As a result the number associated with one of these smaller units may or may not correspond with the overall proportion. Prospective users of survey data from secondary sources therefore need to ensure that, even if small area geographical references were included in the dataset, the units are statistically valid for GIS analysis.

6.2 POPULATION AND OTHER CENSUSES

The earliest records of a census being taken are hidden in the depths of history. However, it is known that counts were undertaken of the population in both the Greek and Roman Empires. Indeed the Roman census being undertaken at the time of Christ's birth is given as the reason why Joseph and Mary travelled to Bethlehem. In the Middle Ages one of the most well-known censuses was commissioned by William I, Conqueror of the Anglo-Saxons, was the Domesday Book. Although this was not a population census in the modern sense of the term in that its principal aim was to identify and record the ownership of property, and may therefore perhaps be regarded as the first known land information system (LIS), the Domesday Survey nonetheless provided some basic demographic data. Modern census taking was signalled 'when enquiries were undertaken solely for the purpose of ascertaining population size' (Dale and Marsh, 1993: 9), rather than for the military conscription or taxation. Modern population censuses in Britain and the United States of America date from the beginning of the 19th century, 1801 in Britain and 1800 in the USA, although France, through its introduction of a system of compulsory registration and the publication of population figures in the 17th century, can perhaps claim to have initiated demographic enumeration.

Housing and population are not the only phenomena to be counted or enumerated by means of national or occasionally regional censuses in the UK and in other countries. If a country's population can be regarded as its most important resource, as the publicity material for the UK's 2001 Population Census suggests, then producing food for that population comes as a close second. It is therefore not surprising that the Agricultural Census dating back to 1866 was the second regular enumeration to be started in the UK. The importance of determining information about food production by means of the Agricultural Census is emphasized by the fact that, although there was no UK Population Census in 1941 due to the Second World War, the annual Agricultural Census was carried on throughout this period and indeed supplemented by the National Farm Survey in England and Wales, 1941–43 (Short and Watkins, 1994). Population and agricultural censuses are carried out in many countries and have a strong geographical emphasis in the collection of the data and the publication of statistical results. These types of census will form the main focus of the remainder of this section, which will examine their characteristics with respect to the UK and other countries. Other forms of census enumeration, such as the breeding bird census will be considered in less detail.

6.2.1 Population Census

The first four censuses (1801 to 1831) in England and Wales were conducted by Overseers of the Poor and by schoolmasters in Scotland and were essentially recorded counts of unoccupied and occupied houses together with tallies of the families and persons present broken down by gender and by a fourfold occupational grouping. The 1841 enumeration was the first census to resemble the current system that employs a schedule (questionnaire) addressed to each householder who completes the required information for each household member. The remaining UK censuses of the 19th and early 20th centuries were relatively limited in their topical coverage and Table 6.1 indicates relatively few additions until 1911. The 1920 *Census Act*, under which even the most recent population census was conducted in the UK on 29 April 2001, outlined the range of topics that should be covered and included the possibility

Table 6.1 Topical content of the Census of Population, England and Wales 1841–2001

Topics relating to housing or household accommodation			
Housing		1841–2001	
Rooms		1901–2001	
Tenure		1961–2001	
Household amenities		1951–2001	
Cars		1971–2001	
Lowest floor level		2001	
Topics relating to persons			
Age		1841–2001	
Sex		1841–2001	
Marital status		1851–2001	
Birthplace		1841–2001	
Relationship within household		1851–2001	
Nationality		1841–1961	
Ethnic group		1991–2001	
Usual residence		1931–2001	
Migration		1961–2001	
Economic position		1871–2001	
Journey to work		1971–2001	
Occupation		1841–2001	
Place of work	1921		1951–2001
Industry		1921–2001	
Size of employer		2001	
Qualifications		1961–2001	
Fertility	1911	1951	1971
Marriage duration	1911	1951	1971
Infirmity		1851–1911	
Limiting long-term illness		1991–2001	
Religion		2001	
Care responsibility (hours)		2001	
Gaelic language (Scotland only)		1881–2001	
Welsh language (Wales only)		1891–2001	

of adding other matters relating to the condition of the civil population. It is this clause that has allowed other topics to be introduced and dropped from subsequent censuses during the 20th and now the 21st centuries. The Household Form for the 2001 UK Census of Population Census is reproduced in the Appendix. In previous UK censuses, although everyone was required to complete the questions applying to individuals, only 10% of forms were processed for the 'difficult to code' topics (e.g. occupation and place of work). However, this division has been abandoned for the 2001 enumeration and all returned forms will be coded, hence all statistical output will be based on the complete enumeration. The 2001 UK Census included more questions addressed to individuals than in 1991, but, despite extensive tests with different forms of wording and categorization, those lobbying for the inclusion of an income question were disappointed when it was decided not to add this topic, mainly for fear that it might depress the level of response and militate the public against the census. Some countries, such as the USA, employ a system of 'long' and 'short' form census schedules in order to deal with questions that take longer to code and the enquiry covers a larger range of topics than in the UK. Details of how the definitions employed in defining topics for the UK Census have changed over the decades may be found in other texts, notably Rhind (1983), Dale and Marsh (1993) and Openshaw (1995), who have edited comprehensive user guides in the first case to the UK Population Census in 1981 and in the latter two instances to 1991.

Some harmonization in the content of censuses carried out in different countries has been promoted by the United Nations' recommendations on the topics to be included in national population and housing censuses. These have been considered within the European context by the Working Group on Population Censuses of the UN Economic Commission in Europe, which has achieved considerable comparability across Europe. There has been success in achieving harmonization in the timing of the national censuses or use of population registration to publish demographic statistics across Europe and other parts of the world. Table 6.2 indicates the population census year in a selection of countries or indicates if a survey or registration is used to produce demographic information. Irrespective of such harmonization, each country's census remains a national event with some countries in the 1980s 'census round' not holding an enumeration due mainly to public dissent and in most cases these have not been reinstated, with equivalent information being obtained from a system of civil registration. Table 6.3 indicates the topics covered in the Population (and Dwellings) Censuses in a selection of countries between 1999 and 2001. Although there is consistency in respect of the core topics (e.g. age, gender, marital status), some countries have clearly expanded beyond the 'standard' United Nations range either to reflect their cultural context (e.g. Scheduled Caste membership in India) or to collect information for specific policy purposes (e.g. time spent on voluntary/unpaid work in the home and elsewhere in Canada and New Zealand). The number of questions column refers to numbered questions addressed to individuals and does not indicate whether these have subsections, hence the table only approximates the burden placed on the population by the different censuses. Canada recorded the highest number of questions amongst the countries listed and also has a population census every 5 years.

Table 6.2 Timing and type of population enumeration for selected countries

Country	Previous date	2000s date	Type of count
Australia	1996	2001	Census
Austria	1991		Census
Belgium	1991		Census
Canada	1996	2001	Census
Czech Republic	1991	2001	Census
Denmark			Register
Finland			Register
France	1990	1999	Census
Germany			Surveys
India	1991	2001	Census
Ireland	1991	2001*	Census
Japan	1995	2000	Census
Luxembourg	1991	2001	Census
Netherlands			Register and survey
New Zealand	1996	2001	Census
Norway	1991		Census and register
Portugal	1991	2001	Census
Singapore	1990	2000	Census
Spain	1991	2001	Census
Switzerland	1991	2000	Census
UK	1991	2001	Census
United States	1990	2000	Census

* The Population Census in the Republic of Ireland was postponed from 29 April 2001 to 28 April 2002 on account of the Foot and Mouth disease outbreak.

The primary purpose of a population census is to count or enumerate the people living within a country or other identifiable area. It is therefore not surprising that geography has always been an important element when undertaking a population census, it forms an organizing framework for the administration of the data collection activity and for the publication of statistical output. One of the crucial roles of geography in the census is that it constitutes a basis for deciding the reference location where people should be counted. Although the census enumeration is tied to a particular day and the date is chosen so as to avoid peak holiday periods or weekdays when people might be absent with work commitments, it is obviously not feasible for the census authorities to require that the population remain at their usual address (geographical location) on census night. It is therefore necessary to devise a population (and household) base for the census statistics. People living at an address can normally be divided into four groups:

- those usually living at the address
- those with another address elsewhere
- those without a usual address
- those staying at the address who usually live elsewhere

Table 6.3 Topical coverage of selected 2000/2001 round population censuses

Country	Year	Questions to individuals	Address on census day/night	Dwelling type	Number of rooms	Number of bedrooms	Tenure	Type of landlord	Rent	Loan repayment	Vehicle availability	Dwelling amenities/condition/material	Dwelling type	Lowest floor level	Number of floors	Non-private dwelling type	Year dwelling built	Year moved to dwelling	Farm residence	Heating fuel	Value	Monthly owner costs	Dwelling condition	Surface area of dwelling
Australia	2001	43	✓	✓		✓	✓	✓	✓	✓	✓		✓			✓								
Canada	2001	51			✓	✓	✓		✓	✓							✓		✓			✓	✓	
France	1999	15	✓		✓											✓		✓						
India	2001	23	✓		✓		✓				✓	✓											✓	
New Zealand	2001	31	✓	✓	✓	✓	✓	✓	✓	✓	✓					✓				✓				
Portugal	2001	24	✓	✓	✓		✓	✓	✓			✓			✓	✓	✓				✓		✓	
Switzerland	2000	21	✓		✓		✓	✓				✓			✓		✓			✓				✓
UK	2001	36	✓	✓	✓		✓	✓			✓	✓	✓	✓		✓				✓				
USA	2000		*		*	*	✓				*	*					*	*	*	*	*	*		

Table 6.3 (*cont.*)

Country	Year	Age	Gender	Marital status	Relationship (family structure)	Birthplace of parents	Nationality/citizenship	Native Australian, Indian, etc. origin	Ethnic Group	Usual place of residence	Previous address 1 or 2 years ago	Previous address 5 or more years ago	Language use	Religion	Highest level of education	Current education/schooling	Educational qualifications	Professional Qualifications	Income	Occupational status/position	Occupation	Industry	Journey to work (time taken)	Mode of travel to work	Distance to place of work	Usual place of work	Size of employer	Hours worked	Fertility
Australia	2001	✓	✓	✓	✓	✓	✓	✓	✓	✓	✓	✓	✓	✓	✓	✓	✓		✓	✓	✓	✓	✓	✓		✓		✓	
Canada	2001	✓	✓	✓	✓	✓	✓		✓	✓	✓	✓	✓	✓	✓	✓	✓		✓	✓	✓	✓		✓		✓		✓	
France	1999	✓	✓	✓	✓		✓			✓		✓				✓				✓	✓	✓	✓	✓		✓			
India	2001	✓	✓	✓	✓				✓				✓	✓	✓	✓				✓	✓	✓	✓	✓	✓				✓
New Zealand	2001	✓	✓	✓	✓					✓			✓	✓	✓			✓	✓	✓	✓	✓		✓		✓		✓	
Portugal	2001	✓	✓	✓	✓					✓	✓	✓			✓	✓	✓			✓	✓	✓	✓	✓		✓	✓	✓	
Switzerland	2000	✓	✓	✓	✓		✓			✓		✓	✓	✓				✓		✓	✓	✓	✓	✓		✓		✓	
UK	2001	✓	✓	✓	✓		✓		✓	✓	✓		✓	✓	✓		✓	✓		✓	✓	✓		✓		✓	✓	✓	✓
USA	2000	✓	✓	*	✓	*	*		✓	*		*	*		*	*	*		*	*	*	*	*			*			

Table 6.3 (*cont.*)

Country	Year	Marriage duration	Infirmity and ill health	Personal computer and internet	Country/place of birth	Literacy status	Place of last residence	Rural/urban status of last residence	Reason for migration	Period of residence since migration	Area of land cultivated	Area of land irrigated	Tenure status of land cultivated	Veteran Status	Providing care to elderly or infirm	Grandparents are carers	Occupational status/position 1 year ago	Ownership of dwelling	Sources of income	Unpaid work and other activities	Immigrant status	Year last in paid work	Language used in house/school/workplace	Weeks worked in 2000	Ancestry
Australia	2001			✓	✓																				
Canada	2001		✓		✓														✓	✓	✓	✓	✓	✓	
France	1999																				✓				
India	2001		✓		✓	✓	✓	✓	✓	✓	✓	✓	✓												
New Zealand	2001		✓	✓														✓	✓	✓					
Portugal	2001		✓	✓	✓																				
Switzerland	2000				✓															✓			✓		✓
UK	2001	✓	✓		✓										✓							✓			
USA	2000				*									*		*	*								

Notes: The census in India includes a preliminary 'houselisting' stage and in Portugal includes a separate schedule with questions about the building. Sample topics in the US census are shown with an *.

The simplest solution is for people to be counted where they happen to be on census night, the persons present base, although this may omit people who are sleeping rough or in transit. Up to 1981 the persons present base was used in the UK Census, although it had become increasingly apparent that changing patterns of activity in society were making this approach less tenable. In 1981 the usually resident population base was introduced, which included residents enumerated in households and communal establishments together with absent residents when at least one person was present in a household. In other words, it did not account for wholly absent households. Further adjustment to the 1981 usually resident population base occurred in 1991 that enabled wholly absent households to be accounted for at their usual address, either if they voluntarily returned a census form, or by imputation where no form was returned or no contact was made by the enumerator. Most output from the 1991 Census used this usually resident base. The notion of the household base is a relatively simple extension of these principles and refers to whether the single person or the group of people comprising a household, by sharing one meal a day or a living/sitting room, were present *in toto* on census night. Although the population and household bases refer to a person's address on census night, these 'places' are administered through geographical areas known as enumeration districts (EDs). These represent 'reasonable' workloads for field staff administering the census forms and have been redefined for successive censuses to reflect the changing distribution of the population and housing at the local scale (Martin, 2000). A significant development in the population base for the 2001 Population Census in the UK is the introduction of a 'one number census' (ONC). This involves enlarging the size of the post-enumeration validation survey to 300 000 households and carrying this out in a structured sample of unit postcodes where under-enumeration is acknowledged as a particular problem. In these areas repeat visits will be made during the coverage survey to produce a total enumeration of households and individuals, which will be used to model the characteristics of missing households and individuals from other postcodes with the same structural features. These imputed counts will be added to information obtained from census returns that pass the initial completion assessment undertaken by field staff in order to produce a final population count. Trials undertaken by the census authorities indicate that this procedure should produce national estimates that are accurate to within 0.13% (ONS/GRO(S)/NISRA, 1999). Some indication of the number of imputed households and individuals will be provided, but for the most part users will be presented with census data as the definitive counts in much the same way as in previous enumerations.

Until the production of Samples of Anonymised Records from the 1991 UK Population Census, the only ways that analysts could get access to census-derived information about households or individuals was through the Longitudinal Study (LS) or by examining historical census returns either in their original paper or in a subsequently created digital electronic format. The LS was started after the 1971 Census and is a 1% sample of persons chosen according to whether their birth date was on one of four days that remain a closely guarded secret within the census offices. Subsequent vital events (e.g. cancer, marriage, death or birth registrations) together with the 1981 and 1991 Census records for these people have been incorporated into the LS, and new people are added who have one of these birth dates either as new births or immigrants registering with the National Health Service. Analysis of these data may be undertaken through a unit based in London Institute of Education, since the digital data records themselves have not left the computer system of the census authority. From a geographical perspective these data have limited value, since they contain only general spatial coding. However, they have proved useful as a means of testing out possible non-standard cross-tabulations that could subsequently be obtained for geographical areas. In the UK and other countries it is possible to examine historical census returns once 100 years have passed since the original enumeration took place. Although such historical census data clearly has a geographical component and may therefore be capable of spatial analysis, a detailed examination of the scope of these data lies beyond the remit of this book.

The standard means of publishing statistics from the UK Population Census from 1801 to 1961 (excluding 1941) was as printed tables of figures in either county reports or thematic volumes. Most of the tables in these publications related to local authority areas (counties and districts), although a limited number of counts (e.g. numbers of male and female residents) and surface area were provided for smaller units (parishes and 'urban' wards). Whether through good fortune or remarkable foresight, the 1920 *Census Act* did not prescribe the types of area for which statistics from the census should be produced. Following the 1951 Census, the first tentative steps taking advantage of this lack of specificity were made when a limited number of census counts were produced for enumeration districts in most conurbations and for some smaller administrative areas. This work was extended after the 1961 Census by a number of studies using enumeration district data mainly for the analysis of urban social patterns (Gittus, 1963–64; Robson, 1969). However, there are three main differences that distinguish the census data used in this early pioneering work and those which were created following the 1971 Census. The early data were only for selected parts of the country rather than nationally; there was no standard format for the tables of counts and they were produced on a bespoke basis for customers instead of being generally available to all potential users.

The first set of standard tables for small areas across the whole country were produced from the 1971 Census, also retrospectively for the 1966 Sample Census, and were soon labelled Small Area Statistics (SAS). This basic concept has continued for the 1981 and 1991 censuses, although the detail in terms of table structure, geography and accessibility have changed both overall and within the constituent countries of Great Britain. The basic idea of an identical set of SAS cross-tabulations across Britain is perhaps a misrepresentation, since from the outset Scotland and Wales had extra tables relating to their additional language and housing (Scotland) questions. Two sets of linked cross-tabulations were produced from the 1991 Census, the Local Base Statistics (LBS) and the SAS, with the former a more detailed version of the latter, but only produced for larger geographical units. Table 6.4 reproduces SAS Table S04 and LBS Table L04 to demonstrate this linkage with a more detailed system of coding used in the latter.

From a geographical perspective the SAS (and the LBS) have undoubtedly become the most important form of output from the UK Population Census, since they have followed or perhaps led the trend in shifting the focus of analysis to ever smaller geographical units and have encouraged wider use of desktop computer mapping and geographical information systems within various user communities. Enumeration districts, which are bounded areas covering the entire country and containing an average of around 225 households, were the obvious smallest units to choose for issuing SAS after the 1971 Census. Conceptually it was a relatively small step for the census offices to regard this type of unit as not only fundamental for census data collection (see above), but also as the spatial building block for constructing and re-aggregating the SAS (and LBS) to larger geographical units. These units have formed the basis of a geographical hierarchy that enables virtually the same set of SAS tables to be produced for the whole of Great Britain down to the smallest ED in England and Wales, or Output Area in Scotland in 1991. Figure 6.1 shows part of this hierarchy emphasizing how the division between the LBS and SAS is related to size of geographical unit and illustrating the number and average population totals occurring in these areas in 1991. Procedures have been applied to blur the counts and suppress the release of statistics according to threshold population and household numbers (50 persons and 16 households for the SAS, and 1000 persons and 320 households for the LBS in 1991) for the purpose of maintaining confidentiality and public support for the census.

In the US the four series of Summary Tape Files (STF 1 to STF 4) in conjunction with the TIGER digital mapping files are probably the type of output from the Population Census that is most suited to GIS analysis. STF 1 and STF 2 relate to the 100% part of the 1990 Census (i.e. the short form questionnaire), with the former covering basic topics, such as age, sex and number of rooms, and the latter contains greater thematic detail. STF 3 and 4 were produced from the long form questionnaire sample data and include data on income, employment and ancestry. Each of the STF series includes at least one file that provides data for small geographical units: STF 1A and STF 3A go down to the block group level and STF 1B to the block level; STF 2A and STF 4A provide data for census tracts, block numbering areas in counties without tracts; and STF 3B relates to 5-digit Zip Codes. Unlike the LBS and SAS produced from the UK Population Census, the STF files contain mostly single counts and means rather than cross-tabulations.

The UK postal geography as an alternative to the standard administrative and ED geography was a very new concept at the time the first set of SAS were produced in the mid-1970s. A UK-wide system of postcodes was introduced in 1971, almost contemporaneously with the 1971 Census. However, the benefits and potential of the postcode system, beyond the immediate improvement of postal collection and delivery, were slow to be realized. The census offices in Scotland took advantage of postal geography as a means of organizing and aggregating census statistics at an earlier stage than in England and Wales. The General Register Office for Scotland (GROS) produced their 1981 SAS for postcode sectors as the second tier in the geographical hierarchy and retrospectively reissued the Scottish 1971 SAS as the Reformatted

Table 6.4 Comparison of Table 4 in 1991 UK Population Census Small Area Statistics and Local Base Statistics. Crown Copyright.

1991 Census Small Area Statistics - 100% Area Identifier - <zoneid> Grid reference - <Easting/Northing>
<...... Ward/Postcode Sector Name... ...> <...............District Name ....> <............County/Region Name ... .>
Table Prefix S04

Table 4 Medical and care establishments: Residents (non-staff) present not in households

Type of establishment	TOTAL PERSONS	Below pensionable age Males	Below pensionable age Females	Pensionable age and over Males	Pensionable age and over Females	Migrants
ALL MEDICAL AND CARE ESTABLISHMENTS	1	2	3	4	5	6
NHS hospitals/homes – pysch.	7	8	9	10	11	12
NHS hospitals/homes – other	13	14	15	16	17	18
Non-NHS hospitals/homes – pysch.	19	20	21	22	23	24
Non-NHS hospitals/homes – other	25	26	27	28	29	30
Housing association homes and hostels	31	32	33	34	35	36
Nursing homes (non-NHS/LA/HA)	37	38	39	40	41	42
Res. homes (non-NHS/LA/HA)	43	44	45	46	47	48

1991 Census Local Base Statistics - 100% Area Identifier - <zoneid> Grid reference - <Easting/Northing>
<...... Ward/Postcode Sector Name... ...> <...............District Name ....> <............County/Region Name ... .>
Table Prefix L04

Table 4 Medical and care establishments: Residents (non-staff) present not in households

Type of establishment	TOTAL PERSONS	Total Males	Total Females	Age 0-15 Males	Age 0-15 Females	Age 16-44 Males	Age 16-44 Females	Age 45 up to PA Males	Age 45 up to PA Females	Age PA-74 Males	Age PA-74 Females
ALL MEDICAL AND CARE ESTABLISHMENTS	1	2	3	4	5	6	7	8	9	10	11
NHS hospitals/homes – pysch.	22	23	24	25	26	27	28	29	30	31	32
NHS hospitals/homes – other	43	44	45	46	47	48	49	50	51	52	53
Non-NHS hospitals/homes – pysch.	64	65	66	67	68	69	70	71	72	73	74
Non-NHS hospitals/homes – other	85	86	87	88	89	90	91	92	93	94	95
Local Authority Homes	106	107	108	109	110	111	112	113	114	115	116
Housing association homes and hostels	127	128	129	130	131	132	133	134	135	136	137
Nursing homes (non-NHS/LA/HA)	148	149	150	151	152	153	154	155	156	157	158
Res. homes (non-NHS/LA/HA)	169	170	171	172	173	174	175	176	177	178	179

Age 75-84 Males	Age 75-84 Females	Age 85 and over Males	Age 85 and over Females	Migrants	With limiting long-term illness	Ethnic Group White	Ethnic Group Black groups	Ethnic Group Indian P'stani and B'deshi	Ethnic Group Chinese and other groups	Type of establishment
12	13	14	15	16	17	18	19	20	21	**ALL MEDICAL AND CARE ESTABLISHMENTS**
33	34	35	36	37	38	39	40	41	42	NHS hospitals/homes – pysch.
54	55	56	57	58	59	60	61	62	63	NHS hospitals/homes – other
75	76	77	78	79	80	81	82	83	84	Non-NHS hospitals/homes – pysch.
96	97	98	99	100	101	102	103	104	105	Non-NHS hospitals/homes – other
117	118	119	120	121	122	123	124	125	126	Local authority homes
138	139	140	141	142	143	144	145	146	147	Housing association homes and hostels
159	160	161	162	163	164	165	166	167	168	Nursing homes (non-NHS/LA/HA)
180	181	182	183	184	185	186	187	188	189	Res. homes (non-NHS/LA/HA)

Notes: PA – Pensionable Age

Tables show cell identification numbers not statistical counts.

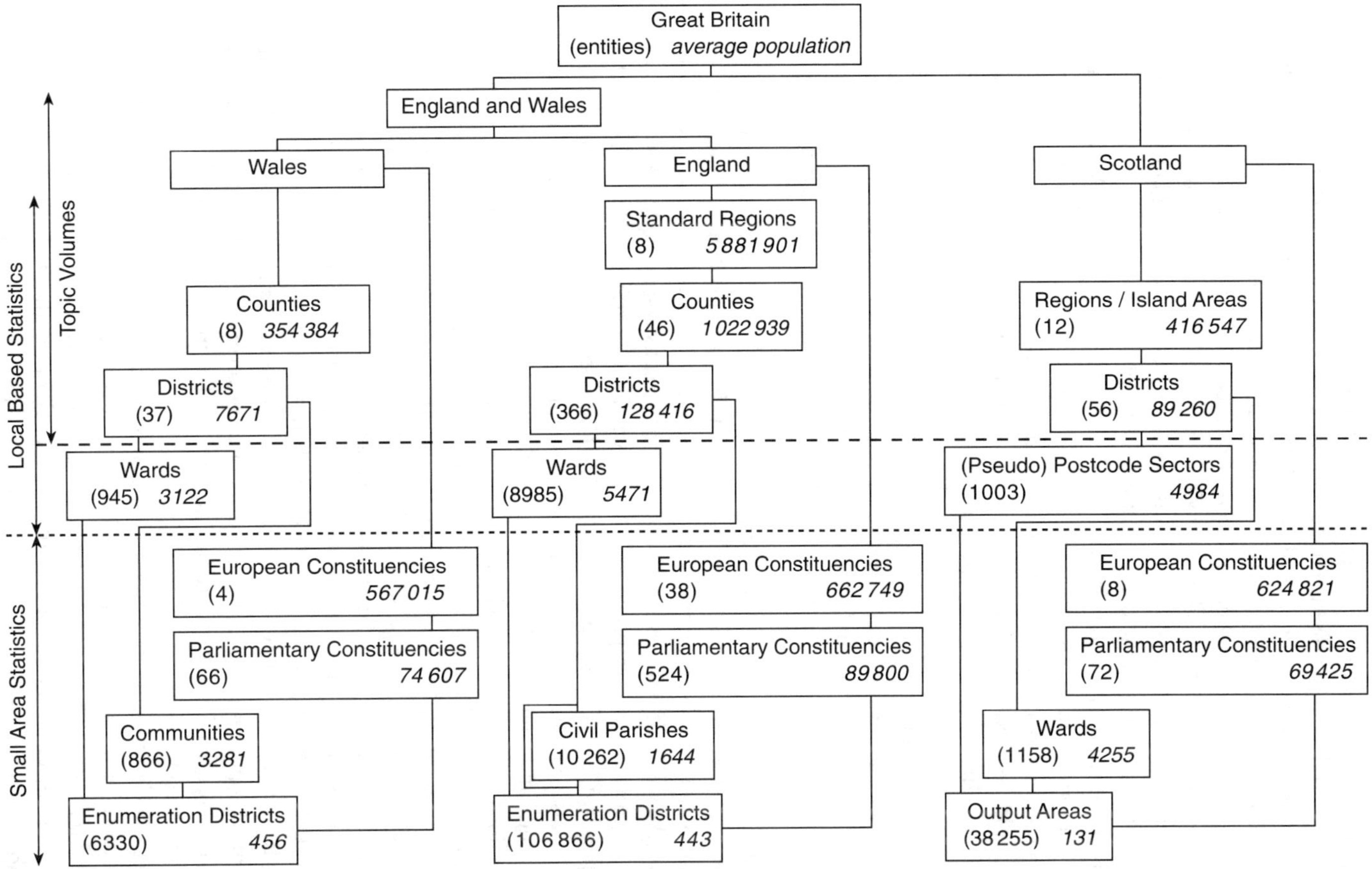

Figure 6.1 Part of the British geographical hierarchy applicable to the 1991 Census of Population.

Small Area Statistics (RSAS) re-aggregated to the 1981 boundary set. These postcode sectors suffered from the disadvantage that some overlapped local government areas and, because of the overriding need to conform to the administrative hierarchy, there were instances of multi-part postcode sectors within the census database. Nevertheless the parts of sectors falling in different districts could be combined to produce complete postcode sectors. The postcode of each address was collected throughout Great Britain in the 1981 and 1991 Censuses, but for addresses in England and Wales it was only entered onto the computer for the first time in 1991.

GROS again went further in 1991 by defining Output Areas (OAs) in terms of combinations of unit postcodes that were not linked to the EDs used for data collection. It was possible for OAs to be smaller than the 'standard ED', since they could be flexibly defined down to the person and household confidentiality thresholds. In the absence of an affirmative decision to use postcodes as a basis for producing statistical outputs from the 1981 Census in England and Wales, Denham (1993) identified the three main purposes for capturing English and Welsh postcodes as: allowing exact statistics to be produced for postcode-defined areas; creating a look-up table linking postcodes and EDs; and facilitating retrospective postcode-based linkage back to the 1991 Census after a subsequent census (e.g. in 2001).

Unlike the UK, where enumerators are employed to deliver census forms, the modern US Census is mailed out by means of the postal system and hence it was necessary to devise a comprehensive address register, known as the Address Coding Guide (ACG). This address register was linked to census geography by means of a street segment-based mapping system known as Geographic Data File – Dual Independent Map Encoding (GDF–DIME). This system was developed for the 1970 US Census and initially covered the metropolitan areas. The GDF–DIME captured the latitude and longitude of the start and end nodes of each street segment, defined as the sections of streets between junctions (intersections), together with the street name. This meant that it was possible to produce relatively simple 'straight line' maps and to create census 'blocks' with which the census forms could be connected by means of the ACG. Subsequently extended to most urban areas for the 1980 US Census, a comprehensive new system known as Topologically Integrated Geographical Encoding and Referencing (TIGER) was introduced for the 1990 Census. Although not completed across the whole of the USA in time for the 1990 enumeration, the TIGER system does allow better maps to be produced and the Census Bureau, through co-operative arrangements with local government bodies throughout the USA, has ensured that the geographic database will be maintained and used for other government surveys.

The US Census Bureau (CB) uses a range of geographical units for the output of population census data, which are divided into two broad categories: legal/administrative areas, which have legally defined boundaries; and statistical areas, which are defined on the basis of criteria prepared by the CB or another federal agency. Figure 6.2 indicates that these areas form a geographical hierarchy in much the same way as the various spatial units in the UK. Blocks are the smallest units for which census data

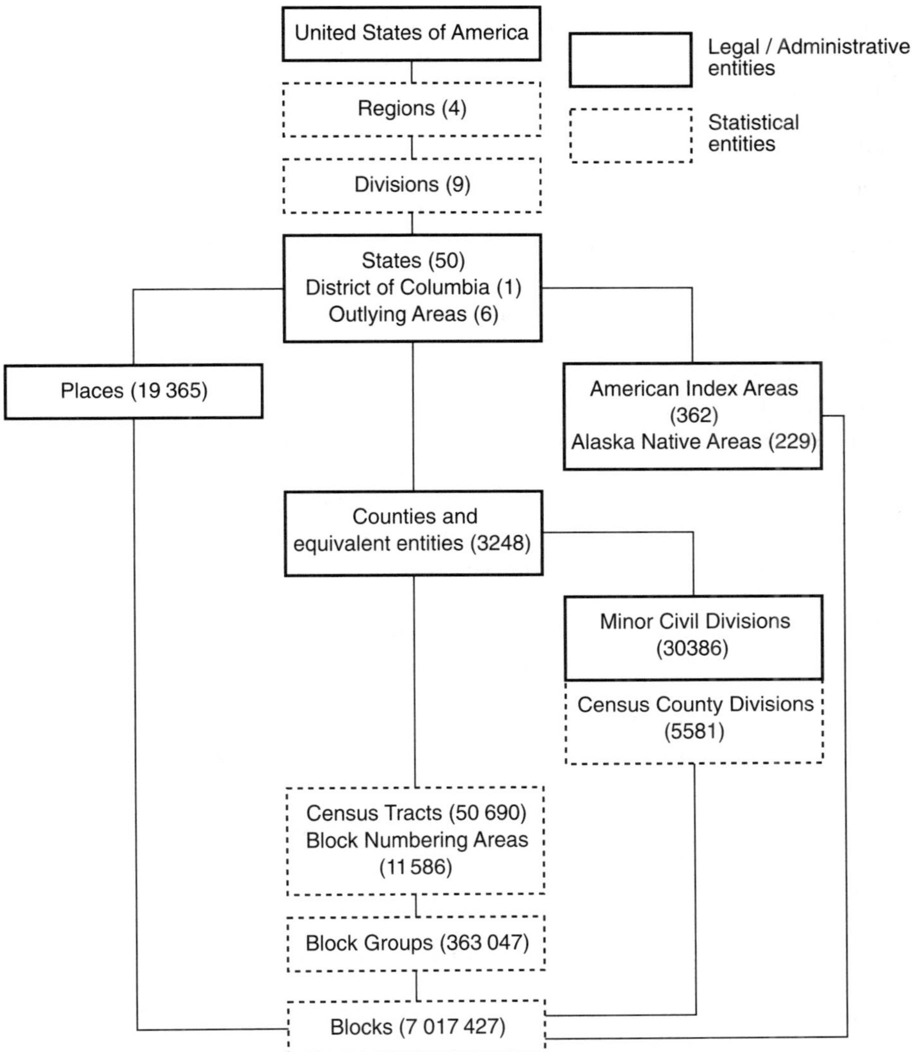

Figure 6.2 Part of the US geographical hierarchy applicable to the 2000 Census of Population.

are tabulated and for the first time the entire USA and its territories was covered by blocks for the 1990 Census producing a total of 7 017 427 units. The five-digit Zip Codes were originally defined for a similar reason to postcodes in the UK, namely to expedite the postal delivery service. As with their UK counterparts, addresses with the same Zip Code frequently cross US State, county, metropolitan area and city boundaries, nevertheless they have become useful for other purposes and are also used as units for reporting the figures from the US Population Census. The primary legal and administrative subdivisions of US States are counties. However, some States and other parts of the USA that lack counties have statistical units that are county equivalents, namely boroughs and census areas in Alaska; parishes in Louisiana, independent cities in Maryland, Missouri, Nevada and Virginia; the part of Yellowstone National Park in Montana; municipios in Puerto Rico; and a variety of units in the outlying areas. The Federal Information Processing Standards Publications (FIPS PUBs) contain details of the hierarchical coding system used to identify the high level geographical entities that are used by the CB and other federal agencies for collecting and reporting statistics not only from the US Population Census but also various other censuses and surveys. Table 6.5 provides small extracts from two of the FIPS Publications for 1990 to illustrate the structure of the coding system.

In the UK a significant break has been made between the geography of data collection and of statistical output for the 2001 Census of Population (Rees, 1997; Martin, 2000). The 2001 EDs have been drawn up using the Ordnance Survey's digital mapping and AddressPoint products which enable addresses to be georeferenced to the nearest metre and postcodes to be checked with householders at the time of census form delivery. Furthermore the capture of address postcodes from the completed census forms and the creation of digital unit postcode boundaries by means of Thiessen tessellation has the potential to allow users to map census output according to postal geography. There has been some debate during the planning period for the 2001 Census about whether a standard set of output areas should be constructed. Figure 6.3 illustrates at a micro-scale ways in which such a set of output areas could be created: one based on aggregations of unit postcodes and the other on address centroids. The problem is that any set would run the risk of being idiosyncratic and incompatible with previous, or even future, censuses, and hence suffer from the same criticisms levelled against using enumeration districts, namely that they are arbitrary and only a proportion remain constant over the intercensal period. Rees and Martin (1997: 28) have commented that 'true compatibility across time can only be established if one boundary set is taken as a starting point'.

At the time of writing the final form of the statistical output has yet to be finalized. In August 2000 the National Statistics office in conjunction with GROS and NISRA produced a draft standard set of statistical cross-tabulations that would comprise the Census Area Statistics (CAS) for OAs. Conceptually these tables are similar to the 1991 SAS and where possible a similar structure has been employed in order to provide some continuity. The proposed tables have been grouped into 11 sections, mainly along thematic lines and including tables that are specific to England and Wales, Scotland and Northern Ireland. Table 6.6, which is based on census authorities draft table outlines, summarizes the proposed content of the CAS. The CAS are planned to contain around some 4800 counts, less than in 1991, but will be enhanced by 300 'Key Statistics' and univariate counts to cover more complex output categories. The change to basing output on 100% of the data collected and the imputation of figures according to the ONC system will introduce discontinuity in comparison with previous censuses. As with the 1981 and 1991 Censuses, other types of output

Table 6.5 Example of US standard geographical entity coding scheme used in 1990 Population Census and other surveys

Metropolitan Statistical Areas

Codes							Titles and Geographical Components
(P)MSA	Level	CMSA	Postal	State	County	Place	
5280	C						Muncie, Indiana
			IN	18			
					035		Delaware county
5320	C						Muskegon, Michigan
			MI	26			
					121		Muskegon county
5350	C	07					Nashua, New Hampshire
			NH	33			
					011		Hillsborough county (part)
						00130	Amherst town
						08100	Brookline town
						37140	Hollis town
						37940	Hudson town
						42260	Litchfield town

1990 Geographic Identification Code Scheme
Counties, County Subdivisions and Places

Geographic Codes				Name	Descriptive Codes					Counts			
State	County	Census Minor Civil	Census Place		Place	Size	Place Description	Standard Metropolitan Statistical Area	Urbanized Area	Tracted	Population	Household	Land Area
42	071			Lancaster County				4000	Part	T			
42	071	005	0050	Adamstown borough pt	00364	04	4	4000		T			
42	071	010	0065	Akron borough	00540	07	4	4000		T			
42	071	015		Bart township	04376	06		4000		T			
42	071	020		Brecknock township	08352	07		4000		T			
42	071	025		Caernarvon township	10704	07		4000		T			
42	071	030	1910	Christiana township	13512	04	4	4000		T			

Source: FIPS PUB 8-5; 1990 Geographic Identification Code Scheme (PHC90-RS).

Figure 6.3 Alternative methods for construction of Standard Output areas in the UK. (a) Centroid georeferenced dwellings, (b) addresses separated in unit postcodes (hypothetical), (c) aggregation according to postal geography, (d) flexible aggregation according to address centroids.

Table 6.6 Summary of proposed Census Area Statistics tables from the 2001 UK Census of Population

Section	Themes	Number of Tables
Univariate	Demographics: Gender, age, migration, living arrangements, dependent children and migration	19*
	Cultural: country of birth, ethnic group and religion	
	Standard occupation: occupation, NS–SEC, Industry	
	Household accommodation: accommodation type, tenure and landlord	
	Health: limiting long-term illness, provision of care and household composition	
	Transport: distance travelled to work and workplace	
Demographic	All people by gender	8
	Age by gender	
	Age by gender by living arrangements	
	Age of FRP and dependent children by gender	
	Age of FRP and dependent children by family type	
	Age by migration	
	Age of HRP and number of depdendent children by migration of households	
	Household composition by migration of households	
Cultural	Gender and age by ethnic group	10
	Country of birth by gender	
	Country of birth by ethnic group	
	Household composition by ethnic group of HRP	
	Age and limiting long-term illness and general health by ethnic group	
	Gender, age and economic activity by ethnic group	
	Industry by ethnic group	
	Tenure and number of cars/vans by ethnic group of HRP	
	NS–SEC by ethnic group	
Health	Gender and age by general health and limiting long-term illness	11
	Tenure and age by general health and limiting long-term illness	
	Gender, amenities and central heating by general health	
	Limiting long-term illness and age by accommodation type	
	Economic activity and hours worked by gender and limiting long-term illness	
	Gender and number of cars/vans by general health	

Table 6.6 (*cont.*)

Section	Themes	Number of Tables
	Age and general health by NS–SEC	
	Gender and age by general health and number of hours worked	
	Gender, economic activity and hours worked by number of hours care provided	
	Households including person with limiting long-term illness and their age by number of carers in household	
Labour Market, NS–SEC, Travel to Work	Gender and age by economic activity	21
	Gender and age by hours worked	
	Economic activity by gender and living arrangements	
	Family composition and number of dependent children by gender and economic activity	
	Gender, age and level of qualifications by economic activity	
	Gender and occupation by age	
	Former occupation by age	
	Gender and industry by age	
	Former industry by age	
	Occupation by industry	
	Occupation and gender by hours worked	
	Economic activity and time since last worked by age	
	NS–SEC by age	
	Gender and NS–SEC by economic activity	
	NS–SEC by household composition	
	NS–SEC of HRP by age of HRP	
	Employed persons and mode of travel to work by car/van availability	
	Gender and age by mode of travel to work	
	Gender and age by distance travelled to work	
	Gender and mode of travel to work by distance travelled to work	
	NS–SEC by mode of travel to work	
Household Accommodation	Number of household space by number of rooms	16
	Accommodation type by household space type	
	Accommodation type and self-containment by tenure and landlord	
	Tenure, landlord and household size by number of rooms	
	Tenure, landlord and persons per room by type of accommodation	
	Household composition by tenure and overcrowding indicator	
	Self-containment, central heating and overcrowding indicator by age	

Table 6.6 (*cont.*)

Section	Themes	Number of Tables
	Accommodation type and central heating by tenure and landlord	
	Self-containment and overcrowding indicator by ethnic group	
	Tenure and amenities by household composition	
	Tenure and lowest floor level by household composition	
	Accommodation type and car/van availability by number of persons 17 or over	
	Household composition by car/van availability	
	Economic activity and age of full-time students by household type	
	Age of students by number of students	
Communal Establishment Accommodation	Communal establishment type by resident type and gender	1
Social Grade	Gender and social grade by age	2
	Age of FRP and dependent children by social grade	
England and Wales Specific Tables	Age by highest level of qualification	5
	Occupation by highest level of qualification	
	Age and highest level of qualification by ethnic group	
	NS–SEC by highest level of qualification	
	Professional qualifications by occupation	
Scotland	Gender and age by knowledge of Gaelic	10
	Gender and distance to place of study by transport mode	
	Age by distance to place of study	
	Age by transport mode to place of study	
	Age by highest level of education achievement	
	Occupation by highest level of education achievement	
	Tenure and whether rented furnished by number of rooms	
	Accommodation type by tenure and whether rented furnished	
	NS–SEC by distance travelled to work	
	Gender by current religion	
Northern Ireland	Age by highest level of qualification	5
	Occupation by highest level of qualification	
	Age and highest level of qualification by ethnic group	
	NS–SEC by highest level of qualification	
	Religion by gender	

Notes: NS–SEC – National Statistics – Socio-economic Classification; HRP – Household Reference Person; FRP – Family Reference Person; *Denotes additional variants within countries of the UK.

will be produced, such as the special collections of migration and workplace statistics, which provide the opportunity to analyse flows of people between areas respectively as a result of having changed their place of residence or for the purpose of travelling to work.

6.2.2 Agricultural Census

The Agricultural Census in the UK is one of a number of surveys undertaken by DEFRA, and by the corresponding government departments in Northern Ireland, Scotland and Wales. As with the population census, it is important that the information collected relates to the situation on all farms on one day in the year. This 'census day' was traditionally 4 June, but was moved to 1 June in the 1970s. DEFRA maintains a register of all agricultural holdings, which is used as the basis for posting out the census forms to farmers, although statistically insignificant holdings are either sampled annually or enumerated on a 3-year cycle. Under the 1979 *Agricultural Statistics Act* (with 1984 amendment) farmers are legally obliged to complete a census return for all the holdings for which they are responsible. Thus, farmers who have acquired farms over the years may have several census forms to complete, even though the holdings may be operated as a single enterprise from a farm management perspective.

The content of the UK Agricultural Census comprised 213 numbered data items in 1998 relating to areas of cropping (to nearest 0.1 ha) and owned/rented land, numbers of livestock and workers. These represent a series of facts about the productive resources on farms, but do not provide any indication of how the products are used. For instance, there is no record of whether cereals are sold off-farm or fed to on-farm livestock, or equally whether milk from cows is processed on-farm into dairy products or sold off-farm. The increased level of detail and specificity in the data items is exemplified by the change from three categories of cattle and two of sheep in the British census of 1870 (Clark, 1982) to 19 of cattle and eight of sheep in 1998. The introduction of new topics is carefully controlled. Fifteen items relating to diversification activities were introduced in 1998. However, 3 years after this event, long awaited by researchers of contemporary agricultural change, the analysis of these data have not been published, since there is some suggestion that some farmers have disregarded this section on the grounds that it is not the concern of DEFRA. This illustrates that care is needed when trying to alter long-established data collection instruments. The accuracy of the returns has also been questioned, for example in respect of whether some entries on the form have been used by farmers as 'balancing items'. There is a suggestion that in some cases the area of grassland has been entered as the difference between total area and the sum of individual cash crop figures.

The US Agricultural Census started in 1840–1841 and continued as a decennial enumeration until 1920, thereafter it has been held quinquennially in years ending in '2' and '7', most recently in 1997. The Canadian Agricultural Census occurs at a similar frequency and agricultural censuses coincide with Canada's Population Census. The content of these North American agricultural censuses is similar to that in the UK, but obviously reflects differences in prevailing crops and livestock enterprises. The 1991 Agricultural Census of Canada included 613 data items, an increase from 311 in 1986, 234 in 1981, 95 in 1976, 240 in 1971 and 130 in 1966. Although the majority of these data items collect information about the areas of agricultural/horticultural land use and cropping and numbers of livestock, in much the same way as the UK's Agricultural Census, it is clear that there has been a greater tendency to introduce other topics. For example, over the period since 1966 farmers have been asked to indicate use of fertilizers, sprays and irrigation, expenditure on selected items, off-farm work, value of agricultural products sold, milk production, land value, mortgages, farmer's age and soil erosion control practices. The Canadian Census Bureau has taken advantage of the coincidence of the country's Agricultural and Population Censuses in 1971, 1981, 1986, 1991 and 1996 to undertake a linkage between the data records derived from the two types of census for a 20% sample of farms, thus enabling the demographic and socio-economic attributes of farm families to be connected with the corresponding farm data (Ehrensaft *et al*, 1984; Shapiro *et al*, 1987). 'Step 8' of the 1991 Population Census of Canada asked 'Does anyone in this household OPERATE a farm, ranch or other agricultural holding?', which provided a mechanism for identifying those households where a link could be made to the agricultural census records.

It is rare, but not unknown, for researchers to gain access to information relating to individual farms that are included in the UK Agricultural Census. This is most likely to occur if undertaking government (DEFRA) sponsored research, but even in these instances it is usually limited to the names and addresses of farmers and possibly the size of holding, which may provide a useful starting point for sampling in micro-scale, behavioural studies. It is extremely unlikely for reasons of confidentiality that researchers will gain access to substantive information from the agricultural census returns. There is not even a comparable ruling applying to the UK Agricultural Census that enables researchers to consult the original returns 100 years after the enumeration, as is the case with the UK Population Census. One notable exception to this rule are the returns relating to the 1941 Agricultural Census which formed part of the National Farm Survey (NFS) conducted between 1941 and 1943. The entire collection of NFS documents, including the 1941 Agricultural Census returns, were made available to the public in 1993 under the 50 years rule and can be consulted in the Public Record Office, Kew (Short and Watkins, 1999). Although not in a digital format, the NFS represents a valuable source of information for historical research and for investigations linking this 1940s information with subsequent surveys of farms (Walford, 1997).

The standard form of output from the agricultural census is published sets of statistics, which divide into two groups – the national/county tables and the parish summaries. During more than a century of the UK Agricultural Census there have been a number of changes in the format of the published volumes of tables, with a separate series for Scotland, and England and Wales as well as a United Kingdom series. The statistical tables comprise total figures for data items on the census form. A subset of these items are included in tables relating to counties, whereas the majority are provided for regions and the constituent countries of the UK. Unlike the output from the population census, relatively little processing of the data by coding the agricultural census questions into frequency counts is undertaken. The main exceptions are in respect of farm size and type, for which standard classifications have been developed, although there are discontinuities, for example when area statistics switched from imperial to metric units (1 acre = 0.405 hectares) and when standard size categorizations based on labour requirements and gross margins were introduced.

The parish summaries have undoubtedly been the most important form of agricultural census output from a geographical perspective. Figure 6.4 reproduces a blank example of the printed form of these Parish Summaries for 1983, which were produced from the start of the Agricultural Census until 1988, with the exceptions of 1868, 1871, 1872, 1892 and 1893. The summaries mainly comprise totals for land use and numbers for livestock and labour, although counts of farms by size and type are also included. The address of each holding is geographically referenced to a single parish irrespective of whether the land connected with a particular holding also lies within the parish boundaries. Since 1949 information relating to a farm has been assigned to the parish containing the majority of its land, whereas previously the allocation had been based on where the farmer lived. The parish summary statistics are simply produced by aggregating the figures for holdings according to their address. Coppock (1960) provided a clear account of the problems resulting from this approach. One potential anomaly is that the physical area of a parish, which obviously includes non-agricultural land, may be less than the summed total area recorded by holdings assigned to that unit. A second issue relates to the compilation of joint summaries for parishes where the total number of farms falls below a confidentiality threshold, three in England and Wales, and five in Scotland. The progressive and geographically uneven reduction in the number of agricultural holdings over the period since the agricultural census started has resulted in the necessity of combining the figures for parishes that had small numbers of holdings. Once combined it is usual for these joint parishes to continue forwards, although this was not invariably the case. Another difficulty relates to the construction of time series statistics. Each holding was assigned a 'county/parish/holding' number with parishes numerically sequenced according to the pre-1974 (pre-1975 in Scotland) local government areas. Although holding numbers and farmer addresses are normally confidential to DEFRA, Table 6.7 illustrates how the county/parish part of this sequence operates and could be translated into the new local government districts that were created by the 1970s reorganization. The parishes are sorted in numerical order (with some omissions evident where past revisions had occurred) and indicates how the numbering produces a seemingly 'chaotic' sequence of both old and new local authorities, since the parishes are also connected with the electoral geography of parliamentary constituencies.

One of the problems with civil parishes (communities in Wales) as a geographical unit is that they varied considerably in size, both in overall area and in the amount of agricultural land. This feature, accompanied by the continued fall in the number of agricultural holdings, led to a reappraisal of the parish summaries system in the early 1990s, following publication of the final set for 1988, because 'it was becoming more and more difficult for MAFF to conceal information about individual farms' (Evans, 1996: 226). MAFF's solution to this problem was to define 'parish groups' for the 1992 census and retrospectively to issue such summaries for 1988 to 1991. The minimum number of farms in these contiguous groups of parishes has been set at 100. From a geographical analysis perspective this change clearly represents a retrogressive step, since these larger areas subsume detailed agricultural land use patterns and limit the researcher's facility for choosing compact areas in which to undertake micro-scale farm surveys.

When carrying out the June 1998 Agricultural Census, MAFF added a grid reference to each of the main (i.e. significant) agricultural holdings in England and retrospectively incorporated these into the databases for holdings in the censuses from 1987 to 1997. These are point grid references to a central field within the holding. Some progress has been made towards adding point georeferences to holdings in Wales, and although incomplete at the time of writing it is envisaged that a comparable collection of datasets will soon be available across England and Wales. This developmental work has enabled the creation of new datasets by allocating farm-level data to grid squares or other polygonal areas, such as Areas of Outstanding Natural Beauty or National Parks, by means of point in polygon manipulation. In some respects this development recollects early work undertaken in the 1960s and 1970s (Scottish Office, 1980) to distribute the parish summaries' data across 1 km grid squares for the Scottish Rural Land Use Information System (RLUIS) and a similar redistribution of the 1971 Population Census statistics (Clarke *et al*, 1980).

However, there is an important difference between the new grid-based datasets and those produced previously for Scotland, namely that spreading parish data involved a certain amount of averaging whereas the new scheme assigns all of the census data connected with an individual farm to either a 5 km or 20 km grid square. The upper part of Figure 6.5a illustrates the case of a parish covering at least half of 17 1 km grid squares and these, with the addition of eight 1 km squares (four from each of two further parishes) make up a single 5 km grid square. The total area of cereals in the complete parish is 113.9 ha and there are a total of 876 cattle. The lower part of Figure 6.5a illustrates how an averaging process assigns 6.7 ha of cereals and 52 head of cattle to each 1 km square and, taking into account the results obtained from similar calculations for the squares covered by the two adjacent parishes, the 5 km square has a total of 164.7 ha of cereals and 1308 cattle. Figure 6.5b illustrates how point grid references to individual farms enable their data to be aggregated to other spatial units, in this example the same 5 km grid square. Since the data for the individual farms identified by crosses was not known, they have been assigned the average value for their parish when calculating the totals of 133.4 ha of cereals and 1191 cattle. Inevitably there remains a certain amount of estimation under the new method partly because holdings may have detached parcels of land that are connected by means of the grid reference to a particular 5 km grid square, when in fact none of that parcel lies within the square. A second difficulty is that holding size has not been taken into account when allocating the data. A 5 km square comprises 2500 ha of land, it would seem quite feasible for this to include 20 or more farm grid references

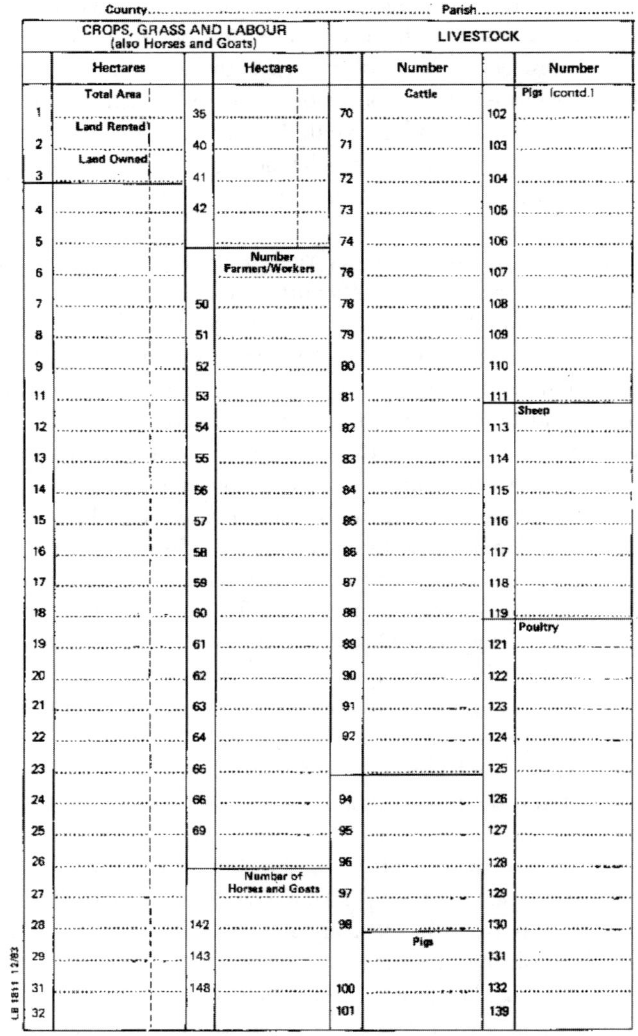

County .... Parish ....

CROPS, GRASS AND LABOUR (also Horses and Goats)				LIVESTOCK			
	Hectares		Hectares		Number		Number
1	Total Area	35		70	Cattle	102	Pigs (contd.)
2	Land Rented	40		71		103	
3	Land Owned	41		72		104	
4		42		73		105	
5				74		106	
6			Number Farmers/Workers	76		107	
7		50		78		108	
8		51		79		109	
9		52		80		110	
11		53		81		111	
12		54		82		113	Sheep
13		55		83		114	
14		56		84		115	
15		57		85		116	
16		58		86		117	
17		59		87		118	
18		60		88		119	
19		61		89		121	Poultry
20		62		90		122	
21		63		91		123	
22		64		92		124	
23		65				125	
24		66		94		126	
25		69		95		127	
26				96		128	
27			Number of Horses and Goats	97		129	
28		142		98		130	
29		143			Pigs	131	
31		148		100		132	
32				101		139	

LB 1811 12/83

Figure 6.4 Example of Parish Summary chart formerly used in England and Wales to tabulate Agricultural Census statistics. Crown Copyright material is reproduced under Class Licence Number CO1W000048 with the permission of the Controller of HMSO and the Queen's Printer for Scotland.

HORTICULTURE

	Hectares		Hectares
	Vegetables		Glasshouse Area
170			
171		205	
			Orchards
172			
173		207	
174		209	
175		210	
178		211	
181		212	
183		214	
184		215	
185		216	
			Small Fruit
186			
187		218	
189		219	
190		220	
192		221	
194		222	
195		223	
196		225	
197		226	
198			
			Hardy Nursery Stock
199			
200		230	
201		231	
		232	
		233	
		234	
		235	
		236	

HORTICULTURE (Cont)

	Number
	Container Grown Plants (Number to nearest 100)
237	
	Hectares Bulbs and Flowers
240	
243	
244	
	Total Horticultural Crops
249	
	Total number of holdings to which this summary relates

ANALYSIS BY STANDARD MAN-DAYS

SIZE GROUPS – S.M.D.'s	Number of Holdings.
1 - 99	
100 - 174	
175 - 249	
250 - 499	
500 - 999	
1000 - 1499	
1500 - 1999	
2000 - 2499	
2500 - 3499	
3500 and over	

A GUIDE KEY IS AVAILABLE

ANALYSIS BY SIZE GROUP

Crops and Grass	Number of Holdings
Nil	
Under 2 ha	
2 ha and under 5	
5 ha and under 10	
10 ha and under 20	
20 ha and under 30	
30 ha and under 40	
40 ha and under 60	
50 ha and under 100	
100 ha and under 200	
200 ha and under 300	
300 ha and under 500	
500 ha and under 700	
700 ha and over	

ANALYSIS BY TYPE OF FARMING

Type of Farming class	Farm Type	Number of Holdings
Specialist Dairy	1	
Mainly Dairy	2	
Livestock rearing and fattening: mostly cattle	3	
Livestock rearing and fattening: mostly sheep	4	
Livestock rearing and fattening: cattle and sheep	5	
Predominantly Poultry	6	
Pigs and Poultry	7	
Cropping: Mostly cereals	8	
General Cropping	9	
Predominantly Vegetables	10	
Predominantly Fruit	11	
General Horticulture	12	
Mixed	13	
Part-time	14	

within some parts of England, as in the example in Figure 6.5b, which may have a total area in excess of 2500 ha.

The addition of grid references to the agricultural census data has also allowed the farm-level information to be allocated to other types of polygonal geography, such as postcode districts and areas, district/unitary authorities and wards and parliamentary constituencies. Confidentiality requirements have necessitated the use of disclosure thresholds, so that areas – grid squares or polygons – containing one or two holdings

Table 6.7 Assignment of parishes in 1985 Agricultural Census to 'Old' and 'New' local authorities in Cornwall

Parish Number(s)	'Old' (pre-1974) local authorities	'New' (post-1974) local authorities
1–2	Kerrier RD	Kerrier
3	Falmouth MB	Kerrier
4	Truro RD	Carrick
5–6	Kerrier RD	Kerrier
7–8	Kerrier RD	Carrick
9–10	Kerrier RD	Kerrier
11	Penryn MB	Carrick
12–13	St Germans RD	Caradon
14	Liskeard RD	Caradon
15–17	St Germans RD	Caradon
18	Liskeard RD	Caradon
19	St Germans RD	Caradon
20	Liskeard RD	Caradon
21–29	Launceston RD	North Cornwall
30	Launceston MB	North Cornwall
31–36	Launceston RD	North Cornwall
37	Kerrier RD	Kerrier
38	Kerrier RD	Penwith
40–41	Kerrier RD	Kerrier
42	West Penwith RD	Penwith
43	St Ives MB	Penwith
44–46	Cambourne Redruth UD	Kerrier
47–48	St Austell RD	Restormel
50	Cambourne Redruth UD	Kerrier
51	St Austell with Fowey MB	Kerrier
52–56	St Austell RD	Restormel
57–62, 64–65	St Germans RD	Caradon
67	Saltash MB	Caradon
68	St Germans RD	Caradon
69	Torpoint UD	Caradon
70–80	Camelford RD	North Cornwall
81	Launceston RD	North Cornwall
82	Camelford RD	North Cornwall
83	Launceston RD	North Cornwall
84	Stratton RD	North Cornwall
85	St Austell RD	Restormel
87	Truro RD	Carrick
89	St Austell RD	Restormel
90	Newquay UD	Restormel
91–92	Wadebridge and Padstow RD	North Cornwall
94–95	St Austell RD	Restormel
96–99	Wadebridge and Padstow RD	North Cornwall
100	Truro RD	Carrick
101	St Austell RD	Restormel
103	Wadebridge and Padstow RD	North Cornwall
105–106, 109–111, 113	Truro RD	Carrick
114	St Austell RD	Restormel

Table 6.7 (*cont.*)

Parish Number(s)	'Old' (pre-1974) local authorities	'New' (post-1974) local authorities
115–116	Truro RD	Carrick
117–122, 124	Stratton RD	North Cornwall
125	Bude-Stratton UD	North Cornwall
126–127	Stratton RD	North Cornwall
128	Wadebridge and Padstow RD	North Cornwall
129	Bodmin MB	North Cornwall
131–134	Wadebridge and Padstow RD	North Cornwall
135	Camelford RD	North Cornwall
136–140, 143	Wadebridge and Padstow RD	North Cornwall
144	St Austell RD	Restormel
145	Lostwithiel MB	Restormel
146, 148	St Austell RD	Restormel
150–155	Liskeard RD	Caradon
156	Liskeard MB	Caradon
157–167, 169	Liskeard RD	Caradon
170–174	Kerrier RD	Kerrier
175	Helston MB	Kerrier
176–186	Kerrier RD	Kerrier
187	Helston MB	Kerrier
188–192	West Penwith RD	Penwith
193	Penzance MB	Penwith
194–196	West Penwith RD	Penwith
197	St Just UD	Penwith
198–202	West Penwith RD	Penwith
203–212	Truro RD	Carrick
213	Kerrier RD	Kerrier
214, 216	West Penwith	Penwith
217	Looe UD	Caradon
218	Truro RD	Carrick
219	Truro MB	Carrick
220	Wadebridge and Padstow RD	North Cornwall
221–222	Launceston RD	North Cornwall

Notes: MB – Metropolitan Borough; RD – Rural District; UD – Urban District.

are suppressed and added to an adjacent unit. Similarly, if one holding accounts for ⩾85% of the total or two holdings for ⩾90%, the data are suppressed. The grid and polygonal datasets include 60 data items from the census including land use areas and livestock totals as well as analyses of holdings by type and size. Despite the possible concerns, this new type of data represent a significant development for the Agricultural Census that will certainly counter the criticisms raised in respect of the parish groups approach.

The US Agricultural Census forms one of a collection of economic censuses and provides data for nearly 3200 counties and their statistical equivalents. Additionally a

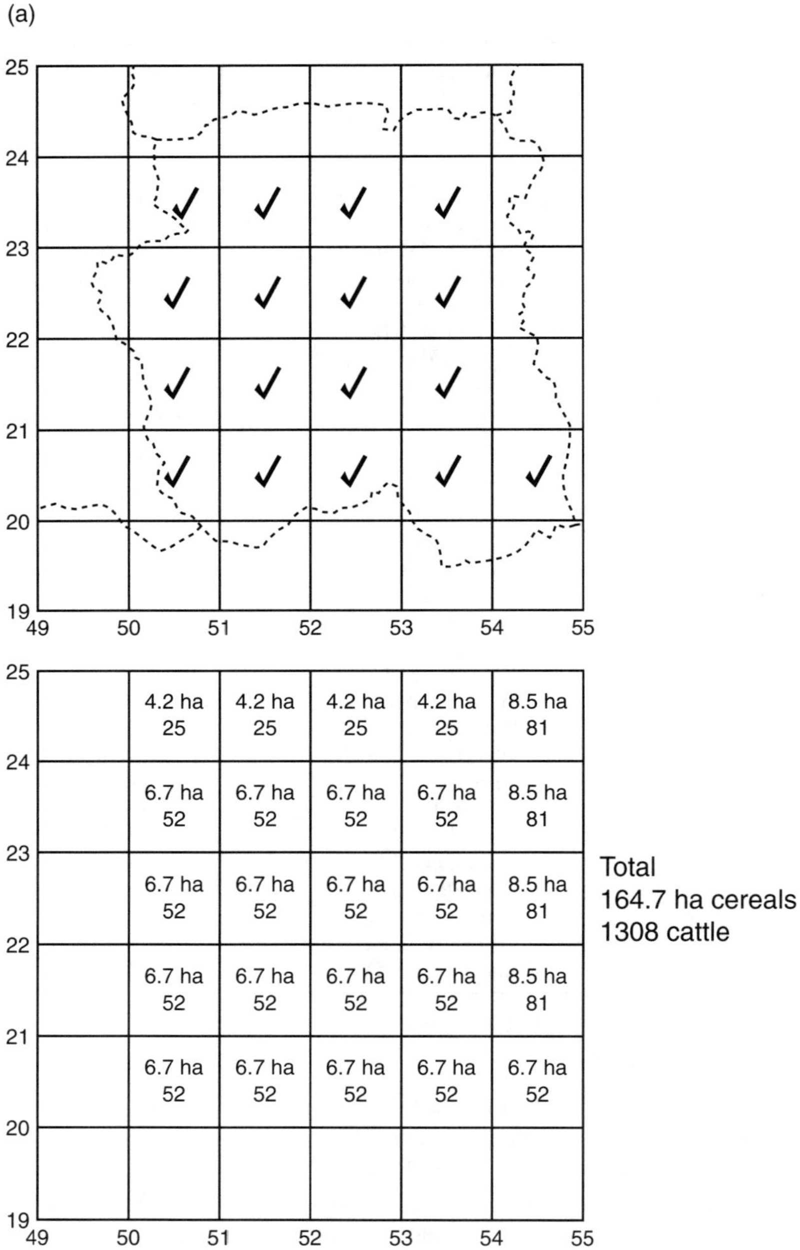

Figure 6.5 Reallocation of agricultural census data to 5 km grid squares. (a) Starting from parish summary statistics, (b) starting from georeferenced farm-level data.

(b)

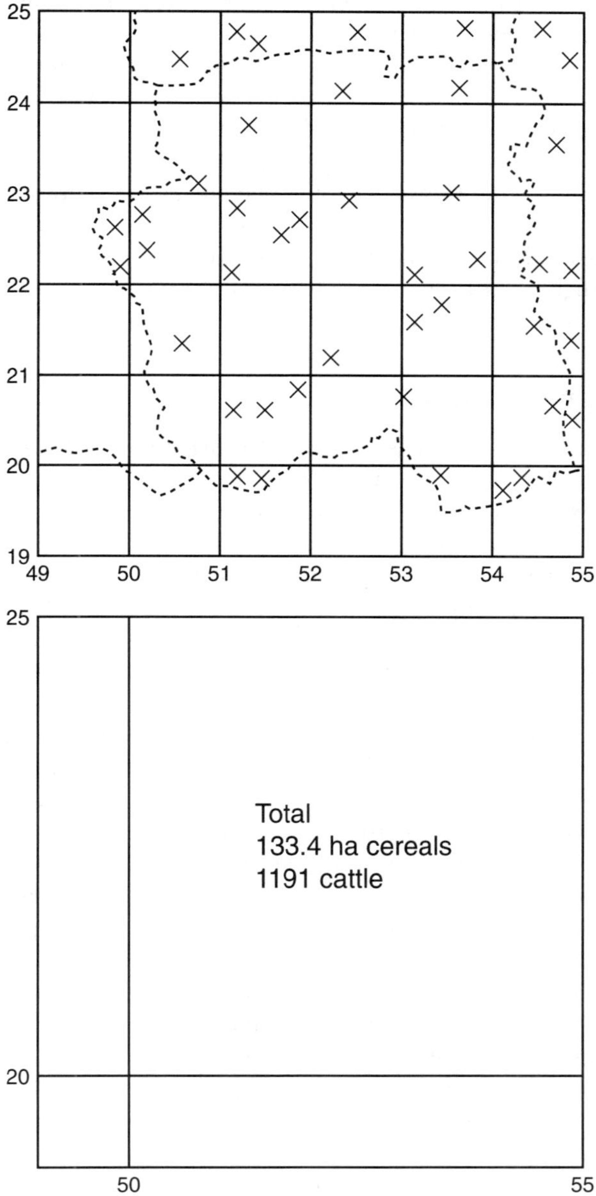
25
24
23
22
21
20
19
49
50
51
52
53
54
55
25
Total
133.4 ha cereals
1191 cattle
20
50
55

limited amount of data were tabulated for five-digit Zip Codes for the first time in 1987. The Agricultural Census of Canada is published for units known as census divisions, which are counties in the east of the country but in effect are simply lines on a map in the west, and these can be aggregated up to large units known as Census Agricultural Regions. In both cases the counties are larger than the traditional parishes that were used in the UK up until 1988 and have tended to be more stable in terms of their individual boundaries and relative position within national geographical hierarchies, although the Census Divisions (counties) and Census Agricultural Regions in Canada are reviewed and in some cases revised every 5 years. Nevertheless, there is still the need for a confidentiality threshold, which is set at 10 farms in the USA. The census figures for counties with fewer farms are combined with a neighbouring unit in much the same way as formerly occurred in Britain.

6.2.3 Breeding Bird Census

Georeferenced counts of biota have been undertaken in many countries for a number of years. In several instances these are referred to as censuses, for example the Common Bird Census in Britain and the Breeding Bird Census in the USA. Changes in the population of different bird species may be indicative of other environmental changes, such as loss of habitats or the presence of pollutants, and the results may be use to inform environmental policy (Crick *et al*, 1998). The 1998 Common Bird Census in Britain was co-ordinated by the British Trust for Ornithology (BTO) and continued a series of surveys monitoring changes in about 70 species that extend back to 1962. Unlike censuses of the human population, where the intention is to record every member of the population present in a country on a particular date, the methodology of bird censuses typically involves repeat visits by volunteers to selected sites that are representative of different habitats, such as farmland, woodland, heathland and wetland, in order to produce estimates of the numbers of individual species (Marchant, 1983). The term census is nevertheless an appropriate one since all contacts with birds are recorded for the date of the field visit. These contacts, including a species identification, are registered on detailed field maps, which are then transferred to individual species maps by the BTO and territories are estimated. A proportion of the 224 sites visited in the 1998 census were either new sites or enumerated by different groups of volunteers and therefore comparison with data for the previous year may be unreliable. Comparable plots between 1997 and 1998 had an average size of some 40 ha, although those on farmland are larger (64 ha), whereas woodland sites are much smaller (16 ha). The long run of Common Bird Censuses makes it possible to monitor change in the incidence of different species and for them to be assigned to categories according to their conservation importance.

The Breeding Bird Census in the USA has a longer history than its British equivalent, dating back to the 1920s. Its co-ordination has recently transferred to the Cornell University Laboratory of Ornithology, which has been responsible for disseminating results in the last decade. As with the British example, volunteers are used to undertake the census. These volunteers choose the location and size of individual plots, which has resulted in some areas of the USA where relatively few counts have been taken and others which have been covered frequently (e.g. in the eastern States). The census uses a spot-mapping procedure (James and Shugart, 1970; Robbins, 1970) to count all singing males and pairs present in a plot on a minimum of eight different days spread throughout the breeding season, from which territory maps can be constructed. The data from US Breeding Bird Censuses of 1990, 1991 and 1993 are freely downloadable over the internet with the intention of encouraging greater use of this under-utilized database, although some long-term analysis has already been undertaken (Johnston, 1990; Johnston and Hagan, 1992). It has to be acknowledged that the way in which the census is organized means that there are questions limiting the potential use of these data for scientific analysis. In particular the lack of control over how sites are selected and the patchiness of the geographical coverage are recognized drawbacks. The data files include some georeferencing including latitude and longitude co-ordinates for the plots and location codes in respect State, county, city or town and the United States Geological Survey map quadrangle.

6.3 SAMPLE SURVEYS

Most sample surveys will include some form of geographical content, even if this is just the fact that they took place in a particular region or country at a given time. As we have seen previously the underlying purpose of a sample survey is to be able to make statements, often of a probabilistic nature, about the parent population from which the sample of observations or phenomena have been selected. In the UK such surveys may be undertaken by official, public sector bodies that range in their scope from national government departments through county, district and parish councils as well as by non-governmental, but still public sector organizations with a regional or local remit (e.g. tourist boards or health authorities). In other words there is decrease in spatial unit size which may be reflected in the geographical specificity of the survey information. Figure 6.6 illustrates that, with a constant sampling fraction of 0.05%, organizations along this spatial scale would be able to achieve a greater level of geographical detail but for a smaller sample size when carrying out some form of household survey. Surveys carried out lower down the scale provide information that relates to a smaller area of geographic space in relation to the location of the individually sampled households, but there are fewer households in the sample and hence it is less representative. Using a constant sampling fraction would mean that 1.5 households were included from the smallest geographical unit (ward). An alternative approach involves increasing the sampling fraction as the spatial unit becomes smaller, which produces better representation. In this example increasing the sampling fraction from 0.05% at the national level through 0.1% at the Government Office Region scale to 10.0% for a survey in an individual ward, would produce more reasonable sample sizes. In other words there is trade off between breadth of geographical coverage and locational specificity.

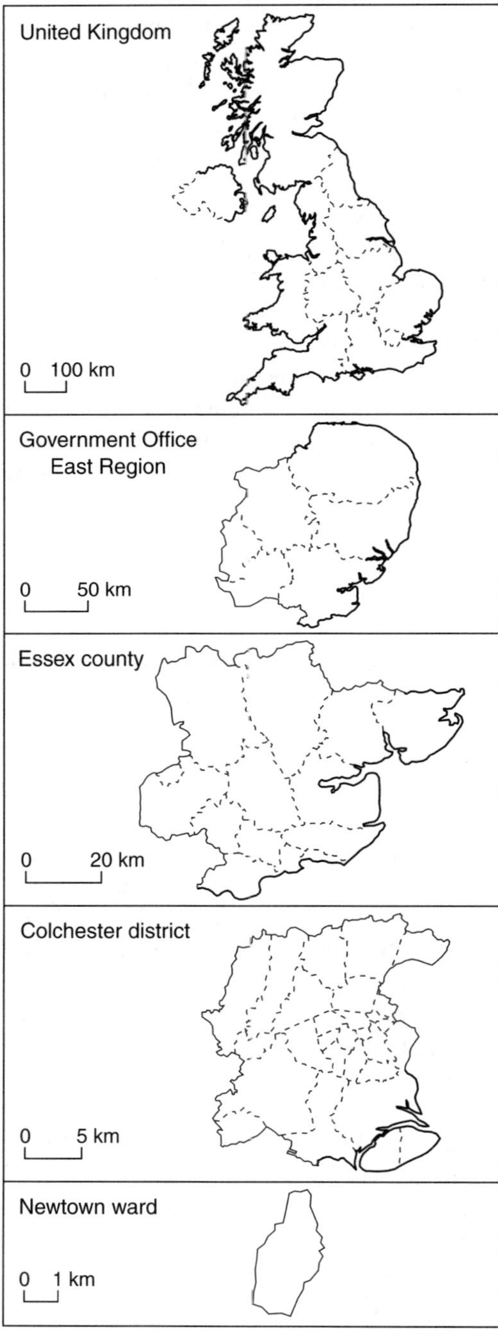

Area	Constant sampling factor (%)	Variable sampling factor (%)	Approximate number of households
United Kingdom	0.05		12 000
United Kingdom		0.05	12 000
Government Office East Region	0.05		1110
Government Office East Region		0.1	2220
Essex county	0.05		220
Essex county		1	4400
Colchester district	0.05		32
Colchester district		5	3200
Newtown ward	0.05		1.5
Newtown ward		10	300

Figure 6.6 Geographical scale and sample surveys.

There are clearly many surveys carried out by organizations at different positions on this spectrum both in the UK and other countries that may or nor contain sufficient individuals to be useful in GIS-based analysis. Two surveys carried out in the UK have been selected for consideration in the following sections to illustrate the issues associated with human and environmental or physical attributes. Both surveys have been undertaken on a number of occasions and are national in geographical coverage, but the issues raised by these examples apply also to one-off surveys and to those carried out in smaller areas.

6.3.1 Countryside Surveys

During the 20th century two major 'censuses' of land use were undertaken in Britain: the First Land Utilisation Survey took place during the 1930s under the leadership of Sir Dudley Stamp, and the field survey for Second Land Utilisation Survey covered the majority of England and Wales from the mid-1960s to early 1970s and was coordinated by Alice Coleman. The focus of both surveys was to produce maps of land utilization in discrete categories for land parcels, they did not attempt to identify ecological habitats or the smaller features of the landscape, such as isolated trees and hedgerows, nor did they attempt to examine variation in basic environmental variables such as soil type and vegetation species. Although these surveys in their own ways represented benchmarks in portraying the distribution of land use and comparisons were made to assess the extent of change that occurred over the interval of approximately 30 years, the intensification of agriculture and pressure on habitats increased the importance and policy relevance of land use research. The Institute of Terrestrial Ecology over the last three decades has developed a series of Countryside Surveys using objective sampling to estimate the stock and change in land use. Association analysis (Williams and Lambert, 1959) was applied to the records of plant and tree species for 2500 woodlands in Britain, which resulted in the definition of 103 groups. The sample for the National Woodlands Survey in 1971 included one representative site selected objectively from each group (plus 34 native pinewoods in Scotland). This established the principle of using stratified random sampling of sites for ecological survey and was the first national classification of a vegetation resource produced using statistical analysis techniques. It provided the quantitative foundation for the series of Countryside Surveys and stimulated the development of new vegetation classification algorithms using correspondence analysis. These were developed into DECORANA (Hill, 1979a) and TWINSPAN (Hill, 1979b), which have been used widely for vegetation analysis throughout the world.

The next stage in the development of this approach was to move from specific woodland vegetation sites to 1 km squares as the sampling unit (Bunce *et al*, 1975) and to measuring a broader range of environmental variables (e.g. altitude and geology) from which a land classification was produced (Bunce and Smith, 1978). The first main trial of this system in Cumbria resulted in the definition of 16 land classes and the classification of some 7000 1 km squares as a result of random selection of squares and vegetation sampling points. On the basis of this work the decision was taken in

1975 to proceed with developing an Ecological Survey of GB with the aim of producing a national land classification. The first Countryside Survey was carried out in 1978 using 256 1 km grid squares each of which contained 11 sample plots and TWINSPAN analysis of the environmental data resulted in the ITE Land Classification comprising 32 classes. The land cover was also mapped in each of these squares with 79 categories of semi-natural vegetation being recognized, from which complete national estimates were made for the whole of Great Britain. The land classes were divided into four main landscapes: arable (2, 3, 4, 9, 11, 12, 14, 25 and 26); pastoral (1, 5, 6, 7, 8, 10, 13, 15, 16 and 27); marginal upland (17, 18, 19, 20, 28 and 31); upland (21, 22, 23, 24, 29, 30 and 32). The second survey took place in 1984 with the grid square sample size increased to 384 and enabled an assessment of change, for example in the length of hedgerows. The data from the two surveys demonstrated that the re-recording of vegetation plots could identify statistically significant environmental change and that integration with satellite imagery allowed the production of land cover maps over large areas. The number of grid squares was increased to 508 for the third Countryside Survey in 1990 and further plots were added along linear features and habitat plots were added in areas of particular conservation interest. A single vegetation classification for Great Britain was produced from all the vegetation data available with 100 individual classes described in terms of several characteristics such as soil types, species number and composition and eight aggregate groups. An important offshoot of the 1990 Countryside Survey was the development of a software application, the Countryside Information System (CIS), to analyse and present the survey data. This provides access to a series of data files georeferenced to 1 km grid squares derived from the Countryside Surveys, for example relating to freshwater invertebrates, geology, hedgerow vegetation and soils. The system also includes files for administrative, electoral and designated areas, such as local authorities, parliamentary constituencies, Areas of Outstanding Natural Beauty and Sites of Special Scientific Interest.

The Countryside Survey 2000 (CS2000) is the most recent in the series with two main components: a field survey of broad habitats, land cover, vegetation, freshwater habitats and linear features in over 500 randomly located 1 km grid squares; and a map and database created from satellite imagery depicting land cover on a field-by-field basis across the whole country. An important new development is that unified information for broad habitat categories across the UK will be produced, since CS2000 coincides with the Northern Ireland Countryside Survey. The CS2000 data now form part of the CIS and thereby provide access to the latest information on environmental change. There has been a clear incremental development of the basic methodology of these surveys over the last three decades and the principle of using stratified sampling and statistical analysis in monitoring land and ecological resources has been emulated in other countries (e.g. Austria, Germany and Spain).

6.3.2 Farm Business Survey

The annual Farm Business Survey (FBS), started in 1936 as the Farm Management Survey, complements the UK Agricultural Census examined in Section 6.2.2 by providing financial data and statistics for a sample of farms. It is paralleled by similar surveys in Northern Ireland and Scotland. In England and Wales the Survey is carried on contract to DEFRA by a group of nine universities and colleges of agriculture on a regional basis. Each of the 'FBS Units' is allocated a quota of farms to make up the overall sample size of approximately 2700. For example, there are about 600 farms in the FBS in Wales. Some farms willingly agree to remain in the survey for several years, up to a maximum of 15, whereas others may only participate once. Contact details of additional farms selected according to stratified random sampling on the basis of size and type are given to the units by DEFRA in order to enlist new farms and thus make up any shortfall in the total.

The content of the survey, which generally involves repeat visits to the farmers, covers physical details of their farm area, cropping and stocking, expenditure on categories of fixed and variable inputs, sales of agricultural and horticultural products, financial data relating to income, capital investments and liabilities, and income from non-agricultural diversification activities. The survey teams are in effect preparing detailed financial accounts for the farms, which are transferred to a central database from which statistics can be produced. This includes weights based on the agricultural census which enables the sample data to be inflated to the national level. In addition the individual units will publish their own 'regional' reports for their areas. Geographical coding in the survey is at the rather general scale of a county to protect the confidentiality of respondents.

The information obtained from the FBS feeds into the European Union's Farm Accountancy Data Network (FADN) to provide a compatible basis for assessing the agricultural industry of the member states. The sample size of the FBS is approximately 2700, while the FADN relates to 60 000 farms. The methodology and content of the individual national surveys is comparable across the EU states. A number of countries carry out similar surveys of farms. In Canada the first Farm Financial Survey took place in 1981 and has been repeated at 2-year intervals since 1988. In the latest survey in 1998 a financial size threshold of $10 000 gross revenue was applied and about 18 000 farms were included across all provinces. In contrast with data derived from censuses, which may be amenable to small-scale spatial analysis, most survey-based data suffer from the generality of their geographical referencing. One of the examples was chosen to illustrate how some surveys that are stratified using small geographical units (i.e. 1 km grid squares in the Countryside Surveys) may well produce nationally useful data that are re-analysed at that scale by people who were not involved with the original data collection. In other cases, particularly where the survey collects information of a confidential or sensitive nature, the scope for detailed spatial analysis may be limited, although it might be entirely feasible to undertake regional or even county scale investigations.

Chapter 7

Administrative and Statutory Data

7.1 CHARACTERISTICS OF ADMINISTRATIVE AND STATUTORY DATA

The previous chapter was concerned with those geographical datasets that are the result of censuses and surveys specifically designed for the purpose of answering research or policy questions. In contrast, this chapter concentrates on those datasets which are either a by-product of an administrative process or generated as a result of a statutory requirement. The former category includes data produced as part of commercial activity, although such data are often difficult or expensive to obtain. It is widely recognized that the need for information becomes greater as the structure and extent of government increases, thus leading in a seemingly inexorable fashion to greater bureaucracy. This process expands the demand for information of all types, ranging from the administration of the fiscal system to the registration of births, marriages and deaths, from recording details about farm pollution incidents to traffic flow on roads. There are a number of similarities between datasets derived from administrative and statutory processes and those collected by means of censuses and surveys. However, before examining the apparent similarities, two crucial differences are considered. Both originate from the fact that data collected by means of censuses and surveys are designed to answer research or investigative questions and their use by people other than the original data collectors is likely to be for similar purposes. The data may in fact have been collected with use as a secondary source of information as an important consideration. In the case of administrative and statutory data, their use is likely to require some form of adaptation before they can be made accessible.

The procedures and instruments used for collecting administrative or statutory data will, by definition, have been designed to conform to a particular set of criteria or objectives. Data collected in these ways may potentially be useful for research or investigative purposes as far as their substantive content and topicality are concerned, but the detailed definitional characteristics of individual datasets may make them difficult to adapt from their original purpose. Furthermore, these, perhaps more than other types of data, are liable to be held under restrictive access arrangements, since, if people and organizations are legally required to provide detailed information about themselves or their activities, it can be argued that they have a right to expect stringent safeguards on anonymity and confidentiality. Most of these administrative and statutory datasets will include the names and addresses of individuals and organizations, and thus these items would need to be removed before the datasets were made accessible as a secondary source of information. Clearly this safeguard would not apply if the public had a statutory right to examine the information, as for example has been historically the case of the electoral roll in the United Kingdom and is contained in Data Protection legislation (Chatto, 1997; Davies, 1997).

The second key difference is that administrative or statutory data are often held in some form of database management system (DBMS) as opposed to a survey analysis package. One of the main reasons is that administrative and statutory data are frequently dynamic in nature, in the sense that they contain information that requires updating and amendment as events occur to the individuals or organizations concerned. In contrast most survey and census type datasets are the result of single data collection events which generate a dataset that is fixed and immutable. The main exception is a panel survey, in which a series of data collection sweeps are carried out with the same respondents over a number of years or even decades in the case of the UK's National Child Development Study, although even in this situation each sweep in a panel survey is a non-continuous event. One consequence of these characteristics of administrative and statutory data is that, particularly if the organization concerned introduced computer-based data handling in the early days of DBMS, they may have been collected in conjunction with customized or bespoke software. This may mean that the data are stored in such a way that there are problems with transfer into more generally available computer systems, although this has become less of a problem as 'off the shelf' solutions to database management grow in popularity.

The similarities between administrative and statutory data and data derived from censuses and surveys relate mainly to the questions of accuracy associated with the use and design of the data collection 'instrument' and with the problems of sampling error, although in this case the error is not the result of sampling itself but occur because of the omission of some observations. In many cases the collection of administrative and statutory data involves the use of a form which is similar to a questionnaire that might be used in a survey. The essential purpose of the form is to act as a means for recording the required information. Current developments in the use of information technology for administrative data processing mean that such data collection instruments are not necessarily printed documents, but may be screen-based representations of a form with facilities to guide the person responsible for collecting the information from the 'respondent' through the various questions and filters. When screen-based forms are employed these would normally add the collected data to an existing database. In the case of printed forms, the data would need to be captured by scanning or manual data entry. Figure 7.1 shows two examples of administrative forms. The first example is of a planning application form (see below), which is used by persons wishing to change the use or undertake the development of a property or parcel of land. Figure 7.1b relates to the electoral registration process and gathers details about the names of people at addresses who are over or expected to reach voting age in order to construct the electoral roll. It is apparent that many of the guidelines relating to the design of a sample survey questionnaire also apply. Clarity and simplicity, which are particularly regarded as hallmarks of good design

(a)

Planning Application

Please read the accompanying Notes for Guidance
Six copies of this form and plans must be submitted
Town and Country Planning Act 1990 as amended

1 **Applicant** (in block capitals)
Name:
Address:
Postcode: Telephone (Daytime):

Agent (if any) to whom correspondence should be sent
Name:
Address:
Postcode: Telephone:

2 **Details of Proposal**
(a) Full address of location of the application site (edged in **RED** on the accompanying plan):
Ordnance Survey grid reference if known:
(b) Describe the proposed development:
(c) Site area: hectares Site dimension (frontage): metres Depth: metres
(d) State whether the applicant owns or controls any adjoining land, and if so, show location edged in **BLUE** on site plan: Yes ☐ No ☐

3 **Preliminary Discussions** - If you have had previous discussions or correspondence with the Council regarding this proposal, please provide details of officer and any reference number quoted by the Council:

I have also applied for Building Regulation Approval: ☐ Give reference if known: /

4 **Type of Application** - Please tick only one large box, and any other small box that applies:
- Full application (including external works, new structure, and/or a change of use): ☐ → The proposal is for change of use only with no external works/new structures: ☐
- Outline application: ☐ → The following matters are included for consideration in this application:
  Siting ☐ External appearance ☐ Design ☐ Means of access ☐ Landscaping ☐
- Reserved Matters: ☐ → Please state the number of the Outline permission: / /
  Please state which of the Reserved Matters are to be considered in this application, if all state 'ALL':
- Removal/variation(s) of condition(s) (including renewal of an existing 'live' permission): ☐ → Please state the number of the permission: / /
  Please state the relevant condition(s):
- If a temporary permission is required, please state the length of period sought:

Figure 7.1 Examples of Administrative Data Collection Record. (a) Planning application form, (b) electoral registration form.

(b)

Don't lose your right to vote

Please read these notes carefully

Note a *Who to include on this form*

- Every adult who normally lives at this address, except anyone listed in Note b. Make sure you include anyone who will be 18 or over by 15 February 2002.
- Anyone who normally lives in your home but is away at the moment (for example, on holiday or in hospital including short-stay voluntary patients in psychiatric hospitals).
- Anyone who is working away from home (for less than six months).
- Students if this is their normal address out of term time.
- Lodgers and anyone else who lives in your home (but not short-stay visitors).
- Merchant seamen who live in your home when not at sea.

Note b *Who not to include*

- **Foreign nationals – only British, Irish, Commonwealth and European Union citizens can register to vote. Do not include any other nationalities.**

Who not to include (continued)

- **Children** born after 15 February 1984.
- **Members of HM Forces, including wives or husbands of HM Forces personnel who have registered as service voters.**
- **People who live or work abroad for more than six months of the year.**
- **Government employees living abroad.**
- **Long-stay patients in psychiatric hospitals.** (Special arrangements can be made for long-stay voluntary patients.)
- **Convicted prisoners who you know will be serving a sentence on 10 October 2000.**

Note c *If no one is entitled to vote*

- If there is no one living in your home who has the right to vote, please write 'No one' in Box 2. Then give the reason (for example, *empty property, foreign nationals, service voters, holiday home*).

Note d *16 and 17 year olds*

- If there is anyone living in your home who was born between 16 February 1983 and 15 February 1984 and who is not listed in Note b, write their name in Box 2 and their date of birth in Box 3. They will become 18 while this register is in force and will be able to vote in any elections after their 18th birthday.

Note e *Over 70 years old (jury service)*

- People aged 70 and over are not called for jury service. So put a cross in Box 4 against the name of anyone who was born before 16 February 1932.

Note f *European Union Citizens*

- If anyone lives in your home who is not a British, Irish or a Commonwealth citizen but is a citizen of another European Union country, put a cross by their name in Box 5. They will automatically be able to vote in local council elections but they will need to apply separately to vote in European Parliamentary elections. We will send them a form to do this.

Please write in BLOCK LETTERS

This form is for the person who lives here now even if it shows the names of someone who used to live here.

If names are pre-printed check the details are correct and, if they are not, change the details to make them correct. Then sign in Box 7.

Box 1. Your full postal address Number of house or flat (or name if not numbered) followed by name of road, then rest of address and postcode.

Box 2. Names of people who should be registered at this address (see notes a, b & c)

Please leave this column blank	Surname BLOCK LETTERS PLEASE	First Names BLOCK LETTERS PLEASE	Box 3 Write the date of birth for 16 and 17 year olds (read note d) Day	Month	Year	Box 4 Put a x next to anyone born before 16 Feb 1932 (read note e)	Box 5 Put a x next to European Union Citizens if other than British or Irish (read note f)

Box 6. Is the building you live in split into flats or bedsits? Please tick Yes or No Yes ☐ No ☐

Box 7. To be signed by the person who has filled in this form.

The information I have given on this form is correct. Signature ______ Date ______

WET THIS STRIP AND STICK IT DOWN

for postal questionnaires, are particularly relevant in the case of administrative forms, especially when they are for completion by the public. The example in Figure 7.1b demonstrates how a minimum amount of information is sought, although it should be noted that experiments have been carried out with using the system of electoral registration to collect a limited amount of additional information by means of electoral enhancement surveys, the intention in these instances being to provide some basic demographic and social data during the decade between full-scale population censuses. Although the purpose for which information is collected differs from that of a census or survey, the manner in which the data gathering takes place is conceptually similar.

The issue of whether data are provided in a voluntary or compulsory fashion might initially seem a source of difference between administrative and survey sources. However, there are examples of data obtained under voluntary and compulsory conditions in each category. For example, most national censuses achieve, or do so very nearly, their 100% coverage by means of making provision of the information a statutory requirement. Some administrative data are collected in a voluntary fashion, although in many cases there is a strong element of compulsion. For example, although it is usually in people's interest to 'sign on' for social welfare payments if they become unemployed or to register as an elector in order to participate in the democratic process, there are some individuals who may for one reason or another decline to take either course of action. Similarly systems for recording pollution incidents may accidentally omit any that remain undetected or unreported. Statutorily required data from individuals, companies or other organizations are, by definition, provided involuntarily. For example, it is required by law that births and deaths be registered and there is an expectation that such vital registration systems operate with 100% effectiveness, however it is not inconceivable that in some extreme circumstances birth or death registration may not occur. Similarly there is a statutory requirement for landowners wishing to change the use of their land or property between one category and another to apply for permission in the case of most uses. Inevitably, there are instances where an application is not lodged with the local planning authority and the change of use might go undetected. These examples illustrate that both administrative and statutory data collection procedures can suffer from the problem of missing observations and thus lead to under- or even over-counting in the datasets.

The incidence of omitted observations from administrative or statutory datasets produces similar statistical problems to those occurring when questionnaires are inaccurately or partially completed, or when sampling is applied. Unfortunately, the sampling error arising from missing data in administrative or statutory datasets may not be random in nature, which is normally assumed to be the case with survey data. This is one of the main reasons why random sampling is recommended for surveys. In administrative or statutory datasets there may be particular types of individual, group or organization which consistently and repeatedly absent themselves from the collection of administrative and statutory data. This deliberate abstension is particularly problematic since data collected for administrative or statutory purposes might initially seem likely to have similar accuracy to a compulsory census in its coverage of the entire statistical population. The apparently authoritative nature of administrative and statutory data might in these circumstances ascribe the data with a perceived level of accuracy that is unwarranted.

Data generated by commercial organizations may be thought of as a particular type of administrative data. It has become increasingly common for both large corporations and small firms to operate some form of loyalty card or product registration scheme that collects information about existing customers. However, this does not provide any information about potential customers, which may be obtained by means of competitions and 'phone-ins'. Figure 7.2 illustrates anonymously a range of such schemes that have occurred in Britain during the 1990s. One of the features of these examples is that they normally include a person's postal geographic code (postcode, Zip Code, etc.). This postal code forms a useful means of linking different data sources together, especially where individual georeferenced addresses within a postcode can be identified. The geographical component of these data has spawned a field of market analysis known as geodemographics (Longley and Clarke, 1995; Martin, 1995; Sleight, 1995). There are several forms of this analysis, which are relevant in commercial and public sector planning arenas, but the most important from a geographical perspective concern branch location, customer profiling and direct marketing. Clearly people working outside or not contractually obligated to an organization seeking to undertake this type of analysis are likely to experience considerable difficulty in obtaining access to this form of data, even for academic research purposes. It is rare for these commercially sensitive data to be available for analysis outside the originating organization, indeed these data may have an in-built bias that would make them unsuitable for academic research.

The nature of administrative and statutory data means that datasets generated from these sources are inevitably more idiosyncratic than censuses and surveys. A substantial proportion of the information collected as a result of administrative and statutory processes, which are associated with central and local government as well as quasi-governmental bodies, is published in the form of statistical abstracts and digests, some of which are now available over the internet. Although not dealing exclusively with such 'official' sources, the following two sections are concerned with information relating to the 'human domain' and to the 'environmental domain'. Much of the information collected as a result of an administrative process in its 'raw' form relates to identifiable individuals or other recognizable units, such as schools, businesses, hospitals and other types of property. This often results in the information only being published in the form of digests and statistical summaries with a fairly general level of geographical coding, typically not going to units smaller than local district or unitary authorities. As of 1 April 2001, Her Majesty's Stationery Office (HMSO) radically changed its procedure for licensing users to access information produced by Crown Agents, such as central government departments. The Cabinet Office Minister, Ian McCartney, announced:

> Under new plans, businesses will be able to utilise and communicate Government information, providing new and exciting products for their customers. Such information includes data about health, house prices, school performance

Thank you for choosing to buy this high quality product from
We are confident that it will give you excellent service. However, in the unlikely event that a problem may occur with this product, we invite you to fill in this registration card to ensure a quick and efficient response.

The information you provide here will be held and processed by Consumerlink. By filling in part A you will enable Panasonic to contact you, if necessary, about your purchase and about new Panasonic products. By filling in parts B and C it is your chance to tell Panasonic about yourself and your needs, which will help Panasonic when designing new products and planning promotional activity. Please feel free to ignore any questions you prefer not to answer.

If you have no objection, your details may be used by Consumerlink and other responsible organisations to contact you with information about themselves and offers, products or services that you may find interesting according to your answers in parts A and C. Please tick this box ☐ if you or your partner would prefer not to share in this opportunity.

Registering your ownership: by filling in these details you will help to contact you, if necessary, about your purchase and about any new products.

1 Please give your name and address :

1 ☐ Mr. 2 ☐ Mrs. 3 ☐ Ms. 4 ☐ Miss

5 Other title (e.g. Dr., Rev., Major)

Your Surname:

First name:

Address:

Town:

County:

Postcode:

Telephone number: 0

E-mail:

Please tell us about yourself: **would also like to know more about you as a person - it helps when designing new services and planning advertising. Knowing more about you also helps Consumerlink and other responsible organisations to ensure that, if you choose to receive information, it will interest you.**

Is the person whose name appears in question 1:

1 ☐ Male? or 2 ☐ Female?

Your date of birth: DAY MONTH YEAR

Partner's date of birth: DAY MONTH YEAR

Marital status:

1 ☐ Married 3 ☐ Divorced/Separated 5 ☐ Living with partner
2 ☐ Widowed 4 ☐ Single

Partner's first name:

Partner's surname:

Partner's Title: 1 ☐ Mr. 2 ☐ Mrs. 3 ☐ Ms. 4 ☐ Miss

5 Other title (e.g. Dr, Rev, Major)

Please indicate the ages of ALL children living at home:

☐ None			
☐ Under 1	☐ 5 yrs	☐ 10 yrs	☐ 15 yrs
☐ 1 yr	☐ 6 yrs	☐ 11 yrs	☐ 16 yrs
☐ 2 yrs	☐ 7 yrs	☐ 12 yrs	☐ 17 yrs
☐ 3 yrs	☐ 8 yrs	☐ 13 yrs	☐ 18 yrs
☐ 4 yrs	☐ 9 yrs	☐ 14 yrs	☐ 19 & over

Thinking about your own home, do you:

1 ☐ Own, or are buying, a house, flat or maisonette?
2 ☐ Rent a private house, flat or maisonette?
3 ☐ Rent a council house, flat or maisonette?
4 ☐ Live with parents/guardians?

(a)

3. PLEASE GIVE A SCORE FROM 0 - 10, 0 BEING THE LOWEST AND 10 THE HIGHEST FOR EACH OF THE FOLLOWING:
(If not applicable eg if the facility is not provided, or if you did not use it, please write N/A)

Welcome ....
Dining Room ....
Breakfast ....
Packed Lunch (If taken) ....
Dinner ....
Service ....
Hospitality ....
Outdoor facilities (e.g. Garden, Pool etc.) ....
Indoor facilities (e.g. Lounge etc.) ....
Value for Money ....
Cleanliness ....
Your stay overall ....

6. ARE THERE ANY ADDITIONAL SERVICES OR FACILITIES AT THIS ACCOMMODATION THAT ARE NOT PROVIDED BUT THAT YOU WOULD LIKE TO SEE?

....

....

....

7. ANY OTHER COMMENTS REGARDING THIS ACCOMMODATION?

....

....

....

Guest's Name: ....

Address: ....

....

.... **Post Code:** ....

(b)

Figure 7.2 Examples of data collection tools used by commercial organizations. (a) Product registration form, (b) visitor survey questionnaire.

> tables, crime, census results and guidance issued by departments. Most of this material can be reused free of charge. Libraries, local government and members of the public will also be able to take out licences. . . . Applying for a licence is simplicity itself. All users have to do is visit HMSO's website, complete a simple application form and hit the submit button. The licence then takes immediate effect (HMSO, 2001).

Although it will remain necessary to protect confidentiality and prevent disclosure of personal information, this development represents a substantial opening up of access that is likely to have important ramifications with respect to analysing and publishing 'official' statistical data.

7.2 DIGESTS AND STATISTICAL SUMMARIES IN THE 'HUMAN DOMAIN'

7.2.1 Planning and Land Use Statistics

The problem of confidentiality is inevitably raised by the fact that much of the information collected as a result of administrative and statutory processes relates to identifiable individuals or units. In some cases rights of public access to this information exist, which means that disclosure of information should not be a problem. A case in point are applications for permission for a change of use or to undertake other types of development of a property that are considered by local planning authorities in the UK. When such planning applications are received the authority has a statutory duty to publicize the proposed development within the vicinity of the site, thereby alerting people to the plans and inviting comment within a specified time period, and to determine the application by reference to development plan policies for the area. The principles of this system were introduced by the 1947 *Town and Country Planning Act* in the pre-computer era and hence created a series of paper records relating to land parcels in the various UK planning authorities. These were eventually superseded by computer-based systems involving database management software to administer and archive the development control process. The potential of these data for research has been recognized for some time, but they were historically difficult to access because of the specificity of software and the lack of a centralized repository for these data. This meant that individual authorities needed to be approached in searching for these data. Summary information at the local authority district level relating to the planning and development system has been published by central government (DETR, 2000a). The current 'Best Value' programme in local government is leading to the development of performance indicators, for example in respect of the percentage of planning applications determined within an 8 week period.

The National Land Use Database (NLUD) is an example of how administrative and statutory data collected by local authorities can be assembled into national information by central government and other agencies. In part it arose out of a project between the Ordnance Survey and the Department of the Environment, initially piloted in 1988, whereby OS surveyors in the field would record changes of land use and these are collated into regional Land Use Change Statistics (DETR, 2001). The NLUD is a partnership between the Department of Local Government, Transport and the Regions, English Partnerships and the Ordnance Survey and has the aim of creating a georeferenced database of vacant and derelict sites and other previously developed land and buildings. It seeks to achieve this by containing comprehensive information about such sites that is freely available to potential developers and other interested groups including researchers. In the mid-1990s the UK government initiated a policy aimed at accommodating a substantial proportion of the forecast future demand for housing requirements on such land. The notional reference date for data collected from local authorities in Phase I was 30 September 1998 with the intention of updating the information in 2001 and annually thereafter. A detailed specification was defined (DETR, 1998) that stated how local authorities were to provide information for the database.

A summary of this procedure provides a clear example of how guidelines from central government specify the format and content of administrative and statutory data. The procedure for collecting the data involved two distinct stages: first, the identification of appropriate sites and the provision of the 'essential data items'; and then 'desirable items' should be provided for selected sites, normally those over 0.4 ha (Table 7.1). The database therefore contains some information about all sites that fulfil the following 'land type' criteria:

- previously developed land now vacant
- vacant buildings
- derelict land and buildings
- other previously developed land or buildings subject to certain conditions

Each site has also been assigned a single land use code using either a specific hierarchical system developed for the NLUD or, for some 'urban' land uses, the Use Classes Order (DoE, 1990). The NLUD classification system has 51 individual classes and 13 divisions (Table 7.2). A detailed procedure was specified for local authorities to follow for identifying sites and assigning their land use category. The extent of this current or former land use and any dependent or closely associated activity was used to delineate the basic spatial units (BSU) in the NLUD by reference to features appearing on the OS Land-Line Plus® plan and each BSU was assigned a unique identification number according to BS7666. This standard defines the information required in order to create a land and property gazetteer relating to Basic Land and Property Units, which are treated as equivalent to BSUs in the NLUD. Georeferencing in the NLUD is provided by assigning a centroid using National Grid co-ordinates to each BSU, although in the first stage the digital boundary co-ordinates were not required. Finally, each local authority completed a summary table that fed into the metadata in the NLUD. This table outlined the completeness, data collection period and source of the site information, and other details relating to the

Table 7.1 'Essential' and 'Desirable' data items collected for the National Land Use Database Phase I

	Essential Data Items Data Item/Question	Instruction/Fixed Response List
1.	Site Reference	*Unique Site Reference Number (automatically generated by database software)*
2.	National Grid Reference	Site centroid: Easting, Northing (*to 1 metre precision*)
3.	Address	*Site Address*
The above fields must comply with British Standard BS7666 which defines spatial referencing and addressing standards for the development of Land and Property Gazetteers (see Annex 5).		
4.	Area (of Basic Spatial Unit)	*Hectares*
5.	Land Type	*Follow step-by-step guide to site identification – see Annex 1* [a] [b] [bz] [c] [d] [e]
6.	Land Use	*Follow step-by-step guide to site identification – see Annex 1* *Use NLUD Version 3.2 Classification or UCOs – see Annex 2 & 3*
7.	Type of Dereliction/Former Dereliction	*Only record if site is derelict (i.e. Land Type [c])* *Use Categories from Derelict Land Survey – see Annex 4*
8.	Owner	[LA] [Other Public] [Private] [Not known]
9.	Planning Status	[Allocated] [Draft Allocation] [Outline PP] [Detailed PP] [None]
10.	Type of Planning Permission or Allocation	[Housing] [Employment] [Retail] [Mixed] [Open Space] [Other] [None]
Local authorities are encouraged to respond to requests 11 and 12 without prejudice to the permitted or allocated use declared at 10.		
11.	Initial Assessment of Suitability for Housing Development	[Suitable for immediate development] [Likely to be developed within 5 years] [Suitable but with Planning or Development Constraints] [Not assessed against local development plan criteria] [Not suitable] [No demand]
12a.	Estimate of Housing Capacity	No. of housing units
12b.		Density per hectare used in calculating 12a
13.	Data Entry on Essential Items	[LA] [EP Surveyor]
Desirable Data Items (fields 15–21 relate to the proposed use of the site)		
14.	Details of Owner	*Only record if Other Public/Private*
15.	Housing Tenure	[Mixed Tenure] [Private] [Social]
16.	Agent	*Development agent for site*
17.	Available on Market	[Yes] [No] [Not known]
18.	Price	£ [Per Hectare] [Per Site Total]
19.	Services Available to Site	[Yes] [No] [Not known]
20.	English Partnership Involvement	[Yes][No][Not known]
21.	Planning/Development Constraints	*Describe constraints on site*

Source: NLUD (1998).

Table 7.2 National Land Use Database classification (version 3.2)

	NLUD Division	NLUD Class	Use Class Order alternatives
1.	Agricultural		Not applicable
		1.1 Field crops	
		1.2 Ploughed fields	
		1.3 Fallow land	
		1.4 Horticulture and orchards	
		1.5 Improved pasture	
		1.6 Field margins	
2.	Woodland		Not applicable
		2.1 Conifer woodland	
		2.2 Mixed woodland	
		2.3 Broadleaved woodland	
		2.4 Undifferentiated young woodland	
		2.5 Scrub	
		2.6 Felled woodland	
		2.7 Land cultivated for afforestation	
3.	Unimproved Grassland and Heathland		Not applicable
		3.1 Unimproved grassland	
		3.2 Heathland	
		3.3 Bracken	
		3.4 Upland mosaics	
4.	Water and Wetland		Not applicable
		4.1 Sea/Estuary	
		4.2 Standing water	
		4.3 Running water	
		4.4 Freshwater marsh	
		4.5 Salt marsh	
		4.6 Bog	
5.	Rock and Coastal Land		Not applicable
		5.1 Inland rock	
		5.2 Coastal rocks and cliffs	
		5.3 Inter-tidal sand and mud	
		5.4 Dunes	
6.	Minerals and Landfill		Not applicable
		6.1 Mineral workings and quarries	
		6.2 Landfill waste disposal	
7.	Recreation		
		7.1 Indoor recreation	D2 Assembly and Leisure
		7.2 Outdoor recreation	(a) A concert hall
		7.3 Allotments	(b) A cinema
			(c) A bingo hall or casino
			(d) A dance hall
			(e) A swimming, skating rink, gymnasium or other area for indoor sports
			(f) An area for outdoor sports

Table 7.2 (*Cont.*)

	NLUD Division	NLUD Class	Use Class Order alternatives
8.	Transport		Not applicable
		8.1 Roads	
		8.2 Public car parks	
		8.3 Railways	
		8.4 Airports	
		8.5 Docks	
9.	Residential		
		9.1 Residential	C3 Dwelling houses
		9.2 Institutional and communal accommodation	C1 Hotels C2 Residential institutions (a) Provision of residential accommodation and care (b) Nursing home (c) Hospital (d) Residential school, college or training centre
10.	Community Buildings		
		10.1 Institutional buildings	D1 Non-residential buildings
		10.2 Educational buildings	(a) for the provision of medical or health services
		10.3 Religious buildings	(b) as a creche, day nursery or day centre (b) for the provision of education (c) for the display of works of art (d) as a museum (e) as a public library or public reading room (f) as a public hall or exhibition hall (g) for public worship or religious instruction
11.	Industrial and Commercial		
		11.1 Industry	A1 Shops
		11.2 Offices	A2 Financial and Professional Services
		11.3 Retailing	A3 Food and Drink
		11.4 Storage and warehousing	B1 Business
		11.5 Utilities	(a) As an office
		11.6 Agricultural buildings	(b) For R&D of products (c) For any industrial process B2 General Industrial B3 Storage and Distribution
12.	Vacant Land and Buildings		Not applicable
		12.1 Vacant land previously developed	
		12.2 Vacant buildings	
		12.3 Derelict land and buildings	
13.	Defence Land and Buildings		Not applicable

Source: NLUD (1998); DoE (1990).

Table 7.3 Summary National Land Use Database information for selected of local planning authorities in England, 1998 (anonymized)

Planning Authority	Previously developed land		Derelict land and buildings		Land occupied by vacant buildings		Land in use with planning allocation for any use/permission for housing	
	Area	Of which proposed for housing	Area	Of which proposed for housing	Area	Of which proposed for housing	Area	Of which proposed for housing
Northern metropolitan	193	43	467	10	47	3	75	3
Northern urban	97	28	148	7	40	9	66	19
Northern rural	12	5	36	2	8	3	1	1
London Borough	4	3	2	—	3	3	8	8
Southern urban	12	6	—	—	23	22	16	5
Southern rural	11	6	310	16	10	4	12	4

Source: DETR (2000b).

anticipated number of dwelling conversions and the average number of dwellings likely to be built on so-called 'windfall' sites. By 2001 a total of 345 local authorities had provided information on 30 000 sites to the NLUD, although only 170 agreed that their data on individual sites could be made available over the internet. Some authorities are concerned because the information is available for free, although this clearly increases the potential of the NLUD as a source of data for researchers let alone for potential 're'-developers of previously developed land. Table 7.3 is based on a supplementary table produced by the Planning and Land Use Statistics Division of DETR and summarizes the type of information available for six unnamed local planning authorities in contrasting areas of England. Although the table does not allow rigorous interpretation of the statistics themselves, it does illustrate the type of information that is available from NLUD. The NLUD has now moved into its second phase with local authorities updating their data to a notional reference date of 31 March 2001.

7.2.2 Population Estimates and Projections

The first population estimates were published by the British central government in 1854 for London, although the current comprehensive system of figures for all local authorities broken down by gender and 5-year age groups dates from 1911 (Cockhead, 1985; General Register Office for Scotland, 1998).The procedure for calculating the population estimates has varied over the years and between the constituent countries of Great Britain. It has now stabilized along the following principles: the home-based all ages population count in a census year is rolled forward to the end of June in that year (mid-year) by adjusting for natural change, migration and other population movements. Then, the census-based age and gender population profile is aged forward to the same date (i.e. by approximately 2 months) allowing for age/gender specific mortality and fertility and migration changes. The counts produced by this process are controlled to match with the all ages figures for each local authority. During the remaining years of a decade the estimates are produced by ageing and adjusting the previous year's figures in respect of age/gender and migration changes, taking into account finalized population census counts and international migration. The latter is accounted for by means of the International Passenger Survey (IPS), which is a continuous sample survey of people departing and arriving at air and sea ports. The population estimates are published annually (Office for National Statistics, 1998) and are also available on electronic media. At present population estimates for areas smaller than local government districts are not produced by the ONS, although a number of local authorities will calculate such figures themselves using one of five methods (Middleton *et al*, 1997) – a local census; cohort survival; apportionment; ratio; and additive change. Simpson *et al* (1996) and Lunn *et al* (1998) concluded, on the basis of examining the small area estimates produced by 41 organizations (mainly local authorities), that local censuses are generally the most accurate approach.

The information used in this process relies on the statutory system of vital registration in which all births and deaths are stored on a centralized computer system known as the National Health Service Central Register (NHSCR). This register represents the system whereby members of the population are issued with and relinquish an NHS number at birth and death respectively. This system also enables an estimate of district to district migration to be determined by recording the event of someone changing their registration with a general practitioner. Although these registration data are used to estimate population totals and for an age group/gender breakdown, aggregate statistics are available on printed and electronic media for wards in the case of births and deaths and for migration flows between government district.

7.2.3 Unemployment Statistics

A substantial proportion of the population will at some time during their 'normal' working age range (say 16 to 60/65 years) for one reason or another find themselves out of work and, if they seek to register for employment, will therefore become a statistic in the monthly tally of the 'unemployed'. Unemployment is not only a personal problem but also a key indicator of economic activity and prosperity that varies regionally and locally. With the transfer of unemployment registration and claimant records from a paper to a computer-based system in 1983 it became possible to augment the monthly count of total unemployment by preparing tabulations of the stocks and flows of males and females broken down by age and duration. These counts, known as the Joint Unemployment and Vacancy Office Statistics (JUVOS), were issued monthly by the Department of Employment in the case of the stock counts whereas the flows were produced quarterly. Initially these were generated at ward level in Britain, which were later accompanied by a postcode sector series. These data formed the core of the National Online Manpower Information Service (NOMIS) based in the University of Durham, which was subsequently extended to include population projections and census statistics, and a facility that allowed users to create maps on a remote computer system and to transfer plot files back over the internet to their local service. The current Nomis UK Labour Market Data service comprises a single distribution centre for labour market data originating from government. The datasets include

- Unemployment Claimant Count
- Census of Employment 1971–1993
- Labour Force Survey annual and quarterly data
- Vacancies handled by Jobcentres
- New Earnings Survey

UK academics can access and download these data for research and teaching purposes subject to defined conditions of use through Nomisweb.

Unemployment benefit, or Jobseekers Allowance, is not the only type of welfare claim available to members of the UK population that constitutes a source of administrative data. The Department of Social Security (DSS) Information Centre provides statistical information from administration relating to claimants of a range of benefits, including Attendance Allowance, Incapacity Benefit, Income Support and Family Credit. For some years this was based on a 5% sample of records, but since 1998 the figures have been based on the complete count and has enabled the DSS small area benefits statistics to be produced. Notable advantages of this system are that the data cover the whole of Great Britain, are up to date and allow linkage overtime by means of the claimants' National Insurance Number. Disadvantages of the system relate to the general problems with administrative data, namely that only the details needed to administer a particular claim are collected and anonymity of claimants needs to be ensured, postcodes are used to aggregate the raw data to wards, although imputation is used for about 0.2% of records, which tend to be geographically clustered, where postcodes are incomplete or invalid. The aggregate statistics include number of claimants, partners and dependents, age group of claimant, household type and gender and relate to a single point in time, currently mid-1998, although mid-1999 were due to be available from March 2001 and mid-2000 from the end of 2001. On the assumption that claiming a benefit indicates a certain measure of disadvantage, these aggregate statistics provide some guidance on where concentrations of social exclusion occur in much the same way as standard socio-economic indicators from the population census are used. The advantage of the claimants data series is that they do provide an indication of relative difference in personal finances, since some benefits are means tested.

7.2.4 Neighbourhood Statistics

It seems as though the unmet demand not only for a nationally produced set of small area population estimates but also social and economic statistics identified by Walford and Hockey (2000) might be fulfilled by a new initiative from ONS, the Neighbourhood Statistics Service (NSS). The service was launched on 26 February 2001 and has been developed as part of the National Strategy for Neighbourhood Renewal and seeks to fill the information gap at the neighbourhood level. At present it is based on bringing together existing datasets containing social and economic information (Table 7.4) that are generated by central government or other organizations. Modification of some of the source datasets has been carried out in order to protect confidentiality, for example university admissions by place of residence and income support claimants have been adjusted to the nearest five using fixed or random rounding. Table 7.4 indicates that the majority of datasets are aggregated to the ward level, although a few are for local authorities, some datasets for electoral divisions in Wales are available and a parallel service is being developed in Scotland to provide additional information. These annually updated datasets can be downloaded in a generic format from the National Statistics Website (http://www.statistics.gov.uk) and thereby analysed using widely available software.

The phased development of the NSS up to 2003/04, when it is expected to be fully

Table 7.4 Data sources in the Neighbourhood Statistics Service

Domain	Dataset title	Source of data
Ward level data		
Economic deprivation	Family credit claimants, March 1999	Department of Social Security, Information Centre
Economic deprivation	Income support claimants, August 1998	Department of Social Security, Information Centre
Education, skills and training	University admissions by place of residence, 1997 and 1998	Oxford University; Universities and Colleges Admissions Service
Education, skills and training	Primary school pupils at Key Stage 2, Summer 1998	Oxford University
Health	Attendance Allowance claimants, May 1998	Department of Social Security, Information Centre
Health	Disability Allowance claimants, August 1998	Department of Social Security, Information Centre
Work	Jobseeker's Allowance claimants, August 1998	Department of Social Security, Information Centre
Work	Incapacity Benefit claimants, December 1998	Department of Social Security, Information Centre
Work	Sever Disablement Allowance claimants, March 1999	Department of Social Security, Information Centre
Work	VAT registered enterprises by industry group, March 2000	Interdepartmental Business Register, Office for National Statistics
Work	VAT registered enterprises by employment sizeband, March 2000	Interdepartmental Business Register, Office for National Statistics
Work	Employee jobs, September 1998	Annual Employment Survey, Office for National Statistics
Work	Claimant count, Oxford University ward level estimates	Joint Unemployment and Vacancy Office Statistics; Oxford University
Indices of deprivation	Indices of deprivation for wards in England, 2000	Department of the Environment, Transport and the Regions
Indices of deprivation	Indices of Deprivation for electoral divisions in Wales, 2000	National Assembly for Wales
Population and Vital Statistics	Oxford University population estimates for wards in England, mid-1998	Oxford University
Population and Vital Statistics	Vital statistics	Population and Vital Statistics Unit, Office for National Statistics
Local authority level – additional or substitute for above ward level data		
Work	Claimant count and claimant count rate	Labour Market Statistics, Office for National Statistics
Population and Vital Statistics	Resident population estimates for local authorities in England, mid-1998	Office for National Statistics; Oxford University

Source: http://www.statistics.gov.uk

operational, includes two important improvements from the geographical data analysis perspective. Although the current aggregate datasets are clearly important for undertaking geographical analysis, further development of the system will also include the precise geographical locations of individual service outlets identified by point georeferences where these are not confidential (e.g. Post Offices) and other infrastructural information. In 2002 a GIS will be incorporated into the system that will provide access to the point georeferenced information. Statistics from the 2001 Census of Population will be incorporated in 2003, which will use the newly defined Census Output Areas as the basic spatial units. It is expected that census counts and a set of key indicators calculated from this enumeration will form a core element in the NSS to which data estimated from existing sources can be attached. The NSS clearly represents a major development in the provision of information about small areas in the UK and demonstrates how administrative sources can be combined with survey and census data to provided a comprehensive resource.

7.3 DIGESTS AND STATISTICS IN THE 'ENVIRONMENTAL DOMAIN'

From a digital data perspective the physical environment is essentially defined by or constructed from a number of the datasets that have been considered elsewhere in this volume, for example, the topographic map and elevation model products available from the OS provide the basic natural and built environmental framework to which various other datasets relate. This section concentrates on a selection of those data sources produced as part of administrative or statutory monitoring processes that specifically concern the measurement of attributes and variables in the physical environment and changes that occur in association with either deliberate or inadvertent human intervention or through non-human (natural) occurrences. It would be impossible to cover all relevant data sources generated within the UK let alone across a range of countries. The intention in this section is to provide some indication of the datasets that are produced as a result of monitoring the physical environment. This monitoring, for example of air, soil and water variables or attributes, is not only important in itself, but these phenomena are part of larger physical processes and their quality has external impacts. For instance, water quality affects the ability of fluvial systems to support aquatic life, hence the monitoring procedures are not only important in themselves, but also have a wider ecological significance. In a number of cases these activities are undertaken by agencies outside of government itself, although frequently public finance is involved with funding the data collection operation. The Environment Agency in England recently commissioned a report concerning the use of digital environmental data (Hooker *et al*, 2000), which covered datasets relating to both the basic spatial infrastructure and to such information as the location of landfill sites, waste treatment plants and river water quality. Further, the environment department of the UK central government, in common with similar bodies in other countries, produces an annual publication that summarizes the current state of the environment, although this tends to be non-specific from a geographical perspective.

7.3.1 Environmental Digest

In the UK the Environment Protection Statistics and Information Management (EPSIM) Division of the former Department of the Environment, Transport and the Regions (now the Department for the Environment, Food and Rural Affairs) publishes an annual *Digest of Environmental Statistics*. Many of the statistics contained in the publication are available elsewhere, but the aim is to assemble figures on environmental protection in a single overarching source. Although the information in this publication is mainly in an aggregate form with only limited spatial coding (e.g. to government office regions), the tables are often based on more detail georeferenced data collected as part of statutory and administrative procedures for monitoring the environment. For example, these include the statutory requirement for local authorities to maintain a register of contaminated land and to maintain records of waste disposal and recycling. The digest covers a series of thematic areas, including air quality, inland water quality and use, coastal and marine waters, and waste and recycling, and the internet version provides links to the raw data used to produce the tabulated information, for instance to the Countryside 2000 Survey reviewed in Chapter 6.

There is a link to the UK National Air Quality Archive which holds air pollution data. The internet site for this Archive (http://www.aeat.co.uk) provides access to the local authority level statistics for 1996 on a grid square basis in the form of internet-based mapping. It should be emphasized that the figures represent estimates of pollutant levels, since monitoring by means of data measurements is only carried out for a series of specific point locations across the country. The estimation algorithm is based on correlations between concentrations of pollutants at specific measurement sites and emission intensity with the results evaluated against the network of monitoring sites (Stedman, 1998). The estimates are regarded as sufficiently reliable to allow modelling of the effect of individual pollutant sources on local air quality. It is also possible to access and download data in comma separated variable (CSV) format from the UK Air Quality Statistics Database at this internet site. This database includes information on 45 pollutants with the most recent figures for 2000 and in some cases with going back to 1970. The system is based on the raw data obtained from the same automatic (hourly measurements) and non-automatic (daily recording) monitoring networks used to produce the mapped output, with the former including sulphur dioxide, nitrogen dioxide, carbon monoxide, ozone (SO_2, NO_2, CO, O_3) and selected hydrocarbons and the former smoke, lead and other trace elements. The periodicity of measurement varies with each pollutant and Table 7.5 indicates a selection of these with their target objective dates. Some pollutants have multiple targets, for example SO_2, which reflect policy aims of reducing air pollution. It is possible to specify individual years or a time series, and to select either rural, urban or all monitoring sites, although the data only give site names rather than georeferences. The output data comprises the aggregate number of periods when the pollutant

Table 7.5 UK Air Quality Statistics Database pollutant periodicity objective targets

Pollutant	Measure	Objective Level	Target date
Sulphur dioxide	No. of periods when the 1 hour mean > 350 μg/m³ (132 ppb)	24 Periods	31 Dec 2004
Sulphur dioxide	No. of periods when the 24 hour mean > 125 μg/m³ (47 ppb)	3 Periods	31 Dec 2004
Sulphur dioxide	No. of periods when the 15 minute mean > 266 μg/m³ (100 ppb)	35 Periods	31 Dec 2005
Nitrogen dioxide	No. of periods when the 1 hour mean > 200 μg/m³ (105 ppb)	18 Periods	31 Dec 2005
Nitrogen dioxide	Annual mean	40 μg/m³ (21 ppb)	31 Dec 2005
Carbon monoxide	No. of periods when the running 8 hour mean > 11.6 mg/m³ (10 ppm)	Zero	31 Dec 2003
Ozone	Provisional objective: No. of periods when the daily maximum of running 8 hour mean > 100 μg/m³ (50 ppb)	10 Periods	31 Dec 2005

Source: UK Air Quality Statistics Database, AEA Technology.

criterion was exceeded and an indication of missing information in the form of the percentage of periods for which data were captured.

7.3.2 River Quality Monitoring Data

The *Digest of Environmental Statistics* also provides a link through to the data sources used for monitoring the quality of rivers. In developed and urbanized countries such as the UK, the chemical composition and flow characteristics of rivers may be strongly influenced by human activities, notably through pollution and abstraction. For example, the discharge of chemicals and other items into water bodies can affect their ability to support wildlife and changes to the nature of the surrounding land surface with the use of various impermeable materials has an impact on the flow characteristics of drainage basins and of rivers themselves. Water quality data are collected by the Environment Agency and by the equivalent body in Scotland, the Scottish Environmental Protection Agency. In England there are approximately 36 000 km of rivers and canals, which are routinely monitored to produce a General Quality Assessment (GQA) and to record particular characteristics.

There are two main aspects of monitoring river quality – biological grading and chemical analysis – and there is a network of sample stretches of rivers where the measurements are made. The biological grading procedure was introduced in 1995 as the start of a 5-year programme. Sampling takes place twice a year (March to May, and September to November) at around 7000 sites on rivers throughout England and Wales with the results from each sweep pooled to take into account seasonality. The grading methodology is based on using macro-invertebrates as indicators of water quality by relating the number found to what would be expected under natural conditions, using a system known as the River Invertebrate Prediction and Classification System (RIVPACS), which was developed by the Institute of Freshwater Ecology (now Centre for Ecology and Hydrology). Only the major types (taxa) of macro-invertebrates are monitored and the number of taxa present are particularly useful indicators of pollution. The Biological Monitoring Working Party has allocated a score to some 80 different taxa and the average score per taxon is recognized as an indicator of organic pollution. The overall biological grading of a river is obtained by calculating two Ecological Quality Indices (EQI), both are proportions that are obtained by dividing the observed into the RIVPACS predicted figures with respect to the number of taxa present and the average scores per taxon. Rivers are then classified on the basis of these EQIs into six groups going from 'very good' to 'bad'.

Chemical analysis of water taken at a series of sample sites is carried out annually on a 3-year rolling programme. The sample sites are usually on the same reaches of the rivers that are used for the biological grading, but they are not necessarily coincident. Samples are taken 12 times a year and the average distance between sample sites is 6 km. The chemical analysis has a number of components, two of which relate to the statutory requirement to monitor levels of ammonia and nitrates, respectively required under the Surface Water Abstraction for Potable Supply Directive and the 1989 *Water Act.* Dissolved and suspended sediments together with biological oxygen demand (BOD) are also included as part of the chemical analysis. The chemical GQA classification is obtained by reference to dissolved oxygen, BOD and ammonia also has six groups from 'very good' to 'bad'.

A similar system of water quality monitoring has operated in New Zealand since 1989, known as the National River Quality Network (Smith and McBride, 1990; Smith *et al*, 1996). There are 77 sample sites on rivers distributed over North and South Islands with 16 measurements recorded every month together with an annual macro-invertebrate assessment. The dataset includes NZ grid co-ordinates for the sample sites and volumetric flow rate, temperature, biochemical oxygen demand, ammonia, nitrate, total nitrogen, dissolved reactive phosphorus, total phosphorus and dissolved oxygen in the set of measured variables. Summary results of the monitoring programme are available from the New Zealand government's statistics system in a similar format to the UK *Digest of Environmental Statistics.* In the New Zealand case the raw data are maintained by the National Institute of Water and Atmospheric Research Ltd. and are accessible from this repository, which parallels the National River Flow Archive that is maintained by the Centre of Ecology and Hydrology in the UK.

7.3.3 Meteorological Data

In many respects the systems of remote sensing of the Earth's surface and atmosphere outlined in Chapter 4 now overshadow ground-based measurement of meteorological variables in the UK. In particular, the geostationary Meteosat operated by European countries produces images of the North Atlantic and Europe every 30 min from 35 780 km above the Equator, while the USA's NOAA polar orbiting satellite at 830 km yields information about cloud structure from its three passes over the same region once at night and once during the day. Nevertheless, there remains a network of some 60 weather stations at which precipitation, temperature and humidity measurements are taken on an hourly basis every day. These recordings can be regarded as administrative data, since the precise range of what is recorded, no more and no less, is prescribed and the information has traditionally been logged on to record sheets with a predetermined structure, and copies of such output can still be obtained. These regular measurements are transmitted to the UK Meteorological Office (MO) and form part of a long-term archive of such data. The MO was founded in 1854 and its origins are in some respects similar to those of the OS insofar as one of the main aims was to provide information for the defence services, in this case naval and merchant seamen. The work of the MO received added impetus during WWI, when it supplied information to air crews and other parts of the armed forces. Although public weather forecasts were rather 'stop–start' initially, they have been regularly published in the press since 1879 with an extension to radio announcements in 1922 and live television broadcasts after WWII, 100 years after the MO's establishment. Modern weather forecasting is based on large-scale modelling of atmospheric processes and depends not only on the specification of the models and the reliability of the data fed into them, but also on having sufficient computing power for 'number-crunching' operations on large quantities of data. The MO has progressively augmented its 'supercomputer' capacity over the last two decades.

A further similarity between the OS and MO in recent years is the change of status to a more commercially orientated organization. Nevertheless, the MO's continuing public service obligations means that it still provides data to researchers as well as the general public. The Website includes access to a rolling 7 day archive of ground-based hourly weather station measurements. These data include wind speed and direction, visibility, temperature, atmospheric pressure level and trend, and weather condition (e.g. rain showers or partly cloudy). In comparison with earlier decades when there was a larger network of recording stations, this represents rather less data than previously, although it is far more accessible and can be downloaded for use in spreadsheet software. Each of the recording stations is named and presented on a regional basis, although there is no form of georeferencing for their location. The British Atmospheric Data Centre, operating from the Rutherford Appleton Laboratory, is supported by the Natural Environment Research Council to provide atmospheric research with access to relevant data, some of which were collected by the MO. In most instances these datasets are not available anonymously over the internet, since verification of research status is necessary. Nevertheless, this does provide an access route to a substantial archive of meteorological data.

Chapter 8

International Comparison of Geographical Data Sources

8.1 INTRODUCTION

The growth in quantity and the generic problems associated with accessing geographical data have already been examined (see Chapter 3). Developing from that discussion and from the preceding examination of the characteristics of the four main sources or types of geographical data, the focus of the present chapter is on specific examples of where digital versions of these data may be obtained and making some comparisons between different countries. It is intended to help readers answer the question 'Where can I get digital geographic data (e.g. satellite imagery) relating to the UK or some other country for my analysis?' However, it would be pointless to attempt a comprehensive and exhaustive listing of such sources on a national scale and at a global one it would be utterly futile. Nevertheless it is worth exploring variations that exist in the availability and characteristics of geographical data across the various countries of the developed and developing world. It is clear that the internet is already acting as a mechanism for accessing geographical data and for carrying out online mapping. This will be examined in Chapter 10 as one of the future directions in which developments in geographical data are likely to move. However, the internet is now also exploited by a growing number of geographical data providers in the commercial and public sector arenas in order to provide information about the datasets that they are able to supply to potential users.

There is obviously little point in replicating in a printed form the information available from the various Web sites managed by these commercial and public sector data providers, although there is merit in assembling a range of details for comparative purposes. The aim of this chapter is to provide an overview of the current state of geographical data provision, although it must be emphasized that changes occur frequently and information soon becomes out of date. In most cases details of the WWW sites refer to the Home Page from which the specific information can be accessed. A range of information sources on printed media and/or the internet have been examined to guide readers to potential sources of useful geographical data. Section 8.2 is intended to allow readers to gain an insight into the general availability of the four types of geographical data and Section 8.3 examines a selection of countries as case studies with a view to attempting to answer the question posed above. It should be emphasized that the details provided are not exhaustive and are indicative of availability, and in many instances a particular dataset may be accessed from places other than those listed.

8.2 AVAILABILITY OF GEOGRAPHICAL DATA

Remotely sensed data, particularly those derived from satellite mounted platforms are international sources of environmental data. Some satellites have been developed with collaboration between countries, others have been placed into orbit by spacecraft operated by a space agency from a different country to the one that developed the satellite. Remotely sensed data obtained by satellites are therefore trans-national and indeed do not explicitly record national political boundaries, unless these happen to coincide with identifiable physical features on the Earth's surface. These characteristics of remotely sensed satellite data together with the worldwide system for receiving such data have resulted in an international rather than a national system for enabling researchers and other potential users to gain access to them.

Aerial photographic surveys are sometimes more local in their scope, partly because they are a more intensive form of data collection and partly because it is feasible for one organization (e.g. a local authority) to commission an aerial photographic survey of its area. Some specialist commercial companies and government defence departments provide an international dimension to aerial photographic survey work, since both types of organization frequently undertake campaigns outside of the 'home' country. Some of the data generated by both satellite and airborne remote sensors have been collected for defence or commercial purposes and, as with other types of geographical data, are subject to restricted access. Nevertheless, it should be noted that even commercially or militarily sensitive data may be released for research some years after they were collected.

This chapter will not examine thematic maps as sources of information for the secondary analyst, on account of their previously identified weaknesses. Attention will focus on the activities and products arising from national mapping agencies and their commercial counterparts as well as intermediate organizations, such as GIS software companies, that also offer pre-formatted topographic mapping (and attribute) datasets for use with their software. Two important aspects of accessing cartographic data will have been apparent from the discussion in previous sections: first, specific sources of such data fall into two categories – analogue and digital maps; and second, the availability of the latter has increased in recent years. The country specific sections of this chapter will concentrate on digital mapping, although in some cases printed editions of these maps will also be published.

The responsibility for conducting censuses and many national surveys frequently lies with central government, although it is equally feasible for a census and other types of survey to be held at a more local scale. Information technology, accompanied by the release of statistics for large numbers of small areas and the growing desire of users to 'get their hands' on the raw data, has had a major impact on the dissemination of census and survey data. In some countries the national census and survey authorities have been willing to relinquish their sole rights to distributing these data and have been willing to utilize other agencies in the academic and commercial sectors provided that adequate safeguards can be implemented. The general spread of information technology has meant that it has become much easier for various types of organization to undertake surveys. Hence, in contrast to the concentration of remotely sensed and cartographic data collection activity within the hands of relatively few organizations, in the case of surveys computer technology has had the opposite effect. It has increased the diversity of organizations undertaking surveys and thereby collecting data that may have some geographical component. The ways in which these have been conducted may make them unsuitable for subsequent analysis, even if the originator was willing for the data to be made available, hence this chapter will concentrate on major surveys.

The previous comments relating to censuses and surveys are also largely applicable to administrative and statutory data. These data are often characterized by a flow between local parts of an organization or institution and the centre. Thus, for example local branches of government or of a commercial corporation may be the initial point of data collection before transfer either immediately or periodically to a central 'headquarters'. One consequence of this from the data access perspective is that there are two corresponding routes to using this type of information for analysis and research purposes. Which is most appropriate depends partly on whether the scale of analysis is local or national. Information in the following sections will be limited to datasets that are collected by national organizations, in particular by central government and its agencies. Detailed investigation by individual researchers will often be rewarded with access to local data that are not widely available.

8.3 UNITED KINGDOM

In the last two decades, but particularly during the 1990s, there has been considerable growth in the availability of digital geographic data in the UK. It is also significant that the originators of many of these datasets have started to adopt a more open policy towards allowing access by users within various communities or groups. Table 8.1 summarizes the position with respect to the existence of geographical data in the four main categories already identified. The principal vendor of remotely sensed imagery obtained from satellite mounted systems is the National Remote Sensing Centre, which is able to supply scenes from the Landsat and SPOT series covering a reasonably long time series. This organization also provides access to digital and analogue aerial photography for a substantial area of the UK through its links with the originator of UKPerspectives. Topographic mapping and elevation modelling products are supplied by a number of organizations, but the Ordnance Survey, having responsibility for maintaining the national topographic map base, is clearly the principal supplier in the marketplace. A comparable range of digital products covering Northern Ireland are available from OSNI including small-scale mapping as well as some thematic mapping. Two other organizations that have been involved with developing digital map products, although generally at a smaller scale than the OS, are the Automobile Association (AA) and Bartholomew, a division of Harper Collins publishers. This scale of mapping has been particularly applicable in respect of transport-related uses, such as route planning and on-board navigation systems. Mapping of geological information has been carried out by the British Geological Survey, Nottingham, with support from the Natural Environment Research Council and has recently developed its DigMap service providing information about the accessibility of digital geological and earth sciences data. At present digital geological mapping for solid formations is complete for the whole of Britain at 1:250 000 and the 1:50 000 scale is in the process of being created from existing paper maps and other sources. In the future three-dimensional data will be added using the BGS Digital Geoscience Data Model.

At a national scale, the various departments of government either directly or through their funding of research investigations play a major part in the development of digital databases of geographically coded attribute data. The important developments in the provision of digital data from selected censuses and surveys have been examined elsewhere in this volume. There is currently a strong emphasis on making access to these data easier, which involves removal of some of the 'red tape' that has previously inhibited users, and cheaper by taking advantage of publishing statistics on the internet. Table 8.1 indicates that there is already national digital coverage in respect of the principal attribute domains across the UK and brings together internet contact points for some of the datasets discussed elsewhere, for example the Agricultural Census, meteorological data and the Neighbourhood Statistics Service.

A number of units have been developed to provide assistance and guidance to potential users of geographical data (Table 8.2). Some of these are 'sector-specific' in the sense that they are aimed at serving particular communities of users, whereas others are more general. The current development of the National Geospatial Data Framework (NGDF) by a consortium of government departments and public sector agencies represents an important step towards harmonizing the approach to data collection and dissemination. It has some similarity to the National Spatial Data Infrastructure in the USA (see below). The initiative was launched at the Association for Geographic Information conference in 1995 and has now moved forward to support the establishment of the UK Standard Geographic Base so that users of geospatial information can benefit from the sharing of data and 'join up' the diverse geographies existing in the UK. An important step towards realizing this aim was

Table 8.1 Selected digital geographical data relating to the UK

Company/Organization	Datasets description	WWW address
Remotely sensed		
National Remote Sensing Centre (NRSC)	Landsat TM multi-spectral, 30 m resolution	http://www.nrsc.co.uk
	SPOT Panchromatic, 10 m resolution	
	Aerial Photographs, 1:10 000; 1:25 000	
Cartographic		
Ordnance Survey (OS)	Topographic mapping, 1:1250; 1:2500; 1:10 000	http://www.ordnancesurvey.gov.uk
	Raster graphics, 1:10 000	
	Elevation models, resolution 10 m	
Ordnance Survey for Northern Ireland (OSNI)	Topographic mapping, 1:1250; 1:2500; 1:50 000; 1:250 000	http://www.osni.gov.uk
	CO-oRdination of Information on the Environment (CORINE), 1:100 000	
	Soil Survey of Northern Ireland, 1:10 000	
	COMputerized Point Address Service (COMPAS), 1 m precision grid references	
	Elevation model, 50 m resolution	
National Remote Sensing Centre (NRSC)	Elevation models, resolution 50 m	http://www.nrsc.co.uk
Bartholomew, Harper Collins Publishers	Topographic mapping, 1:200 000	http://www.bartholomewmaps.com
	Urban areas, 1:50 000; London 1:5000	
British Geological Survey (BGS)	Geological Map data 1:250 000 (1:50 000 in progress)	http://www.bgs.ac.uk
Soil Survey and Land Research Centre	National Soil Map data, resolution 100 m, 1 km and 5 km	http://www.silsoe.cranfield.ac.uk/sslrc
	National Soil Inventory site, profile and elements data, resolution 5 km	

Censuses and Surveys		
National Statistics	Population Census	http://www.statistics.gov.uk
	National Health Service Central Register	
Department of the Environment, Food and Rural Affairs (DEFRA)	Agricultural Census	http://www.defra.gov.uk
	Farm Business Survey	
National Geospatial Data Framework	askGIraffe Data Locator	http://datalocator.askgiraffe.org.uk
	askGIraffe Data Integrator	http://dataintegrator.askgiraffe.org.uk
Administrative and Statutory		
Department of Transport, Local Government and the Regions (DETLR)	National Land Use Database	http://www.dtlr.gov.uk
		http://nlud.org.uk
Department of the Environment, Food and Rural Affairs (DEFRA)	Digest of Environmental Statistics	http://www.defra.gov.uk
National Statistics	Neighbourhood Statistics Service	http://www.statistics.gov.uk
Meteorological Office	Meteorological data	http://www.metoffice.gov.uk

Table 8.2 Selected academic/research organizations providing access services to digital geographical data in the UK

Organization	Datasets and Services	WWW address
Data Archive, University of Essex	4000+ datasets including national and local surveys, agricultural and population censuses, economic, health, political and social social surveys.	http://www.data-archive.essex.ac.uk
	On-line catalogue	http://biron.essex.ac.uk/cgi-bin/biron/
Edinburgh Information and Access (EDINA), University of Edinburgh	Digimap (OS digital maps)	http://www.edina.ac.uk
	UKBORDERS™	
Manchester Information and Associated Services (MIMAS), University of Manchester	Map datasets – Bartholomew digital maps data and Census boundary data	http://www.mimas.ac.uk
	Population datasets – Censuses of 1981 and 1991, and Vital Statistics	
	Spatial data – Landsat and SPOT satellite data	
	Survey datasets – Farm Business Survey, Family Expenditure Survey, Family Resources Survey, General Household Survey, Health Survey for England, Labour Force Survey, British Household Panel Survey, National Child Development Study and 1970 British Cohort Study	
Centre for Ecology and Hydrology	National Water Archive	http://www.ceh-nerc.ac.uk
	Environmental Information Centre	
	Biological Records Centre	
	Countryside Information Service	http://www.cis-web.org.uk
Rutherford Appleton Laboratory	British Atmospheric Data Centre	http://www.rl.ac.uk

achieved by the development of the *askGIraffe Data Locator* and the *askGIraffe Data Integrator*. The Data Locator is a system for finding out about the existence of geospatial datasets by using a keyword or map-based geographical search procedure. Users are able to examine the metadata relating to the datasets, including their availability, and in some instances to view samples. The Data Integrator is intended as a way for users to obtain access to definitive sources of data for administrative, electoral and postal geographies that conform to the UK Standard Geographic Base. It also provides detailed information about the nature of these geographies and terminology employed in respect of geographical data.

The academic community in the UK has been well served by a number of organizations that have been established since the mid-1960s with the aim of preserving and providing access to data. The UK Data Archive, as it is now known, based at the University of Essex, the first of these units, was established in 1967 and had the explicit aim of archiving and cataloguing the (mainly survey) data and documentation that UK academic researchers were creating. From these innovative, but perhaps fairly humble beginnings the Data Archive has now assembled a collection of over 4000 datasets not only produced by academic researchers but also by central and local government, national agencies and commercial companies (e.g. National Opinion Polls). There was never any intention to concentrate exclusively on georeferenced data, although clearly most surveys have a geographical component. Nevertheless, the Data Archive includes within its holdings small area data from the Population and Agricultural Censuses, election study datasets and various specific localized surveys. The Data Archive has pioneered the development of a standard system for cataloguing digital datasets that is comparably rigorous with systems developed for printed publications (see Table 8.2 for WWW address).

Two other UK university-based data centres that have developed some specialization in the provision of georeferenced data are EDINA at the University of Edinburgh and MIMAS at the University of Manchester. Two services at EDINA are particularly worth highlighting in the context of this book. The first is UKBORDERS™ that contains a growing collection of boundary files relating to historical and current legally constituted units of government and areas defined specifically for statistical purposes (e.g. census enumeration districts). EDINA also operates the Digimap service that provides the UK academic community with access to a range of OS digital data products, including the Land-Line® data, for teaching and research purposes. The WWW pages of Edinburgh University's Geography Department (http://www.geo.ed.ac.uk/home/giswww.html) also includes a comprehensive links section to internet sites related to GIS, which includes some locations where spatial information can be obtained. Although MIMAS (formerly MIDAS) has included some spatial data within its portfolio of datasets, for example Bartholomew digital maps, digital boundaries for the 1991 Population Census and satellite imagery from the Landsat and SPOT series, its main focus has been on attribute datasets for which there is wide national interest within the academic community. MIMAS and associated units at the University of Manchester have developed a specialization in Population Census area and anonymized records data. A notable example is the

Casweb system, which provides direct internet-based access to the aggregated area statistics from the 1991 Population Census. This has recently been extended to enable users to download formatted digital boundary data corresponding to the areas selected. Further internet-based access tools have recently been developed under an initiative funded by the Economic and Social Research Council and Joint Information Systems Committee (ESRC/JISC) to facilitate easier access to previously underused population census statistics, to allow redistricting of data between different boundary sets and matching of counts between the 1971, 1981 and 1991 censuses. The Vital Statistics for Wards are also available for on-line access from MIMAS for 1981 onwards.

8.4 FRANCE

Since 1938 the IGN has conducted a regular programme of vertical aerial photographic surveys to produce analogue maps. The current programme has a 5 year average repeat cycle for monochrome and colour photography at a scale of 1 : 25 000 for *départements* (Table 8.3). In total over 4 000 000 individual photographs have been taken since 1921 that are available to users. The collection of analogue imagery is currently being augmented by the use of digital photography which helps to speed up the interval between the time of the flight and when the information becomes available, and provides more opportunity to process the images using computer-based systems. The series of black and white aerial photographs at a scale of 1 : 30 000 provides the basic data of IGN's topographic mapping, which contrasts with the UK where the OS has traditionally employed field survey as the favoured means of data collection in a smaller territory. The existence of *SPOTImage*, which was reviewed in Chapter 4, has given France a leading role in the development of satellite-based remote sensing systems in comparison with other countries in western Europe.

IGN has developed a number of digital mapping databases for use in GIS in recent years. There are two main topographic map products, one containing all topographic elements and the other with the data divided into nine themes, including, for example, administrative units, buildings and railways. These databases have been created by combining aerial photographic and field survey data and include a digital terrain model with a resolution of 25 m and an accuracy of 2 m created by interpolation from contours and spot heights. Another more general terrain model has been created by capturing the contours from the 1 : 25 000 and 1 : 50 000 topographic maps combined with aerial photography. IGN has also derived digital land use mapping at scales from 1 : 50 000 to 1 : 500 000 and contain information relating to seven themes. Georeferenced raster map files at a range of scales from 1 : 25 000 upwards are also available and have been produced from scanned images of the paper maps.

The French equivalent of the National Statistics Office in the UK is coordinated by the *Conseil National de l'Information Statistique* (CNIS) and provided through the *Insitut National de la Statistique et des Études Économiques* (INSEE) in conjunction with a number

Table 8.3 Selected digital geographical data relating to France

Company/Organization	Datasets description	WWW address
Remotely sensed		
SPOT Image		http://www.spot.com http://sirius.spotimage.com
Institut Géografique National (IGN)	Aerial photographs, resolution 1:25 000; 1:30 000	http://www.ign.fr
Cartographic		
Institut Géografique National (IGN)	Topographic mapping, 1:5000; 1:10 000 and 1:25 000 Elevation models, resolution 75 m	http://www.ign.fr
Censuses and Surveys		
Administrative and Statutory		
The Institut National de la Statistique et des Études Économiques (INSEE)	Population Census Agriculture Education Environment Health Industry Infrastructure and transport Work and employment	http://www.insee.fr

of government departments and agencies. INSEE is responsible for conducting the Census of Population in France, the most recent of which took place in March 1999, and a number of other regular surveys. In addition INSEE also plays a key role in the decentralized French statistics system, which is organized around a series of themes or domains (agriculture, education, environment, health, industry, infrastructure and transport, work and employment, etc.). These can be accessed via the INSEE internet side (Table 8.3), which in turn provides hypertext links to other sites that form part of the French public statistics system. As an example, the environment domain is operated by the *Institut Français de L'Environment* (IFEN) providing details of relevant sources of data. This catalogue can be viewed either thematically or through a map-based interface. Selecting an individual region (e.g. *Nord – Pas de Calais*) produces the following list of topics:

- Fauna and flora
- Water
- Climate and air quality
- Soil
- Built environment and infrastructure
- Waste and pollution emission
- Health risks
- Economic activity
- Politics and society

Selecting waste and pollution emission as the topic produces a list of information sources relating to different categories of waste and pollutant emission including waste from hospitals, industry, local authorities and of radioactive origin and liquid, noise and solid pollutant emissions into the environment. Although details of the interface are different, there is considerable similarity between this metadata source and the *askGIraffe* system that has been developed in the UK. The aim in both cases is to provide a means of connecting users who have a question about the availability of certain information with details of the organizations responsible for collecting the raw data. Although this represents an important step forward, in many of these systems the inquiry will reveal that the data are not accessible, or only at a cost, outside of the originating organization.

8.5 UNITED STATES OF AMERICA

It is undoubtedly the case that users of GIS wishing to access digital geographic data relating to the USA have been able to do so for longer and more easily than in many other countries. In part this relates to citizens' rights of access to information enshrined in the American Constitution and in part to the earlier drive in the USA to convert data into a digital format. The central involvement of the USA in space missions and in pioneering satellite remote sensing clearly means that there is a well-established infrastructure providing access to these data. One of the main sources of geographical data is the United States Geological Survey (USGS), which has aerial photographs covering the USA since the 1950s and worldwide satellite imagery since the early 1970s to the present. The USGS Earth Resources Observation System (EROS) Data Centre provides a centralized ordering service, while the Global Land Information System (GLIS) comprises an interactive internet-based service that contains 'metadata' or descriptive information about datasets relating to the Earth's surface.

The US government started the National Aerial Photographic Program (NAPP) in 1987 with the aim of photographing the land area in either colour or black and white, or in some localities both emulsions, every 7 years. Information about availability and progress in this programme can be viewed on the USGS internet site (see Table 8.4). The original aerial photography programme was later developed to start producing digital orthophoto quadrangles (DOQs) from 1993 with the aim of covering the coterminous USA by 2002. The intention is to update these on a 10 year cycle and at 5 years in areas where change is more rapid. The USGS is also the primary source of topographic mapping and elevation data for the USA and provides coverage at the scales indicated in Table 8.4. Cartographic data are compiled from various sources including aerial photographs, legal documents and field survey. The digital line graph (DLG) data files have been derived from the topographic quadrangle maps and most of the USA is covered at the 1 : 20 000, 1 : 24 000 and 1 : 25 000 scales, although some areas are at the smaller 1 : 62 500 scale. These are vector data files and are available in nine themes in a similar fashion to the IGN system in France (see above). Some of the smaller scale DLGs (1 : 100 000 and 1 : 2 000 000) are available over the internet and via file transfer at no cost to the user. The USGS website includes details of how these digital files can be obtained.

Attribute data for geographical entities, typically counties or county statistical equivalents, are produced from the US Census of Population and Housing and the various economic censuses in a digital form, on computer tape, CD-ROM or over the internet (see Table 8.4). In some cases, notably the Censuses of Housing and Population, Retail Trade, Service Industries and Manufacturers, data are also produced for smaller geographic units (e.g. ZIP Codes). There is a range of regular national surveys, where the level of geographical detail is quite general, although in some particular cases the information is statistically reliable at the county and equivalent scale.

The National Spatial Data Infrastructure (NSDI) (National Academy of Sciences, 1995) in the USA arose out of recognition that many organizations using GIS share a common need for data in seven domains:

- elevation and bathymetry – a spatially referenced position above or below a datum surface
- hydrography – the georeferenced location and major features of lakes, ponds, streams, rivers, canals, oceans and shorelines
- geodetic control – a common reference system for establishing the co-ordinate positions of all geographic data

Table 8.4 Selected digital geographical data relating to the United States of America

Company/Organization	Description	WWW address
Remotely sensed	Satellite imagery	
EarthWatch		http://www.digitalglobe.com
Orbimage		http://www.orbimage.com
Space Imaging		http://www.spaceimage.com
NASA		http://edcwww.cr.usgs.gov
		http://eos-am.gsfc.nasa.gov
United States Geological Survey (USGS)	Satellite imagery	http://edcwww.cr.usga.gov/webglis
	Aerial photographs 1 : 40 000	http://www.usgs.gov/www/products/status.html
Cartographic		
United States Geological Survey (USGS)	Elevation models, 10 m and 30 m	http://www.usgs.gov
	Digital Orthophoto Quadrangle 1 m and 8 m resolution	
	Digital Line Graphs, 1 : 24 000	
	Raster graphics, 10 m resolution	
	Land Use and Land Cover, 1 : 100 000 (partial); 1 : 250 000	
Censuses and Surveys		
Bureau of Census	Census of Agriculture	http://www.census.gov.
	Census of Manufacturers	
	Census of Population and Housing	
	Census of Retail Trade	
	Census of Service Industries	
	Current Population Survey	
	American Housing Survey	
	County Business Patterns	
	Current Business Reports	
	Annual Survey of Manufacturing/Current Industrial Reports	

	Current Construction Reports Agricultural Economics and Land Ownership Survey Farm and Ranch Irrigation Survey	
Administrative and Statutory		
Federal Geographic Data Committee (USGS)	National Spatial Data Infrastructure	http://www.fgdc.gov.us

- cadastre – the spatial information necessary to represent and describe the geographic extent of past, present and future rights and interests in real property
- transportation – the georeferenced location and major features of roads, trails, railroads, waterways, air and sea ports and bridges and tunnels
- governmental units – the georeferenced areas of the legal units of government identified according to the Federal Information Processing Standard Code
- orthoimagery – a georeferenced image of the Earth prepared from aerial photography or other remotely sensed data

The NSDI is administered through the Federal Geographic Data Committee (FDGC) within the USGS and has brought together federal, regional, state and local agencies, private companies and utilities to build and maintain this commonly used 'framework' of geographical data. One of the underlying aims of the project has been to create a common basis and standard for geographical data that can be used by GIS analysts in this wide range of organizations. It has been important to develop the principal of sharing data and information between organizations and to this end the National Geospatial Data Clearinghouse provides a means for users to find, evaluate and access geographical data through the internet.

8.6 AUSTRALIA AND NEW ZEALAND

The Australia Land Information Group (AuSLIG) comprises the National Mapping Agency and has developed a number of digital data products specifically for use in GIS (Table 8.5). There is national coverage at 1 : 250 000 scale in raster format available on CD-ROM including viewing software. AuSLIG offers two digital vector map products at the same scale, TOPO-250K Series 1 has national coverage but fewer themes than Series 2, which covers hydrography, infrastructure, relief, vegetation and reserve areas. A series of digital elevation models are available, although some, particularly at a smaller resolution, do not at present have national coverage.

New Zealand's equivalent organization is the National Topographic/Hydrographic Authority of Land Information (LINZ). The Topographic Map 260 Series at a scale of 1 : 50 000 have been converted into a digital format called NZ*Topo*. As with other digital vector map products outlined with respect to various countries in this chapter, data are organized into a series of eight themes – hydrography, relief, utilities, transport, non-vegetative land cover, structures, vegetative land cover and named features. The point, line and polygon objects within these themes are named as object classes, for example, the hydrography theme includes springs (spring_pnt), rivers (river_cl) and lakes (lake_bdy, lake_poly and lake_seed). The suffixes for the object classes denote whether the digital co-ordinates refer to a point (pnt), line centre or edge (cl or edg) or an area boundary, polygon or seed point (bdy, poly and seed). LINZ has also developed a sophisticated land information system known as Land*online*, which is a programme designed to convert cadastral records that the New Zealand government has accumulated over a period of more than 100 years into a digital format. The conversion process covers 3 million title records and 2 million title instruments, 1.2 million plans and 1.3 million parcels and 40 000 geodetic survey marks. The land parcel framework in this system replaces the Digital Cadastral Database. Access to this service requires user registration and incurs charges. A comprehensive archive of aerial photography commissioned by New Zealand government departments and agencies dating back to 1937 is held under Crown Copyright by NZ Aerial Mapping Ltd.

The Australian Bureau of Statistics (ABS) is responsible for conducting the Population Census in Australia which is held quinquennially in years ending '1' and '6', with the latest enumeration having taken place on 7 August 2001. A comprehensive internet site (see Table 8.5) provides information about the Australian Census of Population, including a table indicating the topics that have been asked in each enumeration since 1911, and a range of other surveys and censuses conducted by the Bureau. For example, the Survey of Employment Arrangements and Superannuation was conducted with about 25 000 respondents throughout Australia between April and June 2000. The ABS has produced a wide range of 'Confidentialized' Unit Record Files (i.e. anonymized data files) from many of its surveys, including the 1981, 1986, 1991 and 1996 Censuses of Population and Housing. The 1996 reference file represents a 1% sample of the full unit record base with the categorizations of some variables collapsed in order to preserve confidentiality. Statistical output at a fairly general geographical scale can be viewed at the WWW site and there are details of products and procedures for ordering data on electronic media.

The operations of Statistics New Zealand (SNZ) are governed by the 1975 *Statistics Act*, which charge this department not only with the tasks of collecting and assembling statistics across a wide range of thematic areas and informing policy in these topical fields, but also of advising the Minister of Statistics of the New Zealand government on statistical policy. While there are parallels with some other countries considered in this chapter, there is clearly a more centralized approach, with SNZ playing an important consultative and advisory role in the data collection and survey activities of other government departments. The datasets covered by SNZ include those collected by means of specific regular or one-off surveys as well as others derived from administrative and statutory procedures. Table 8.5 indicates some of the census, survey and administrative datasets covered in the metadata listings that can be viewed at the SNZ website, referring in particular to those having more detailed geographical coding.

In common with the UK and USA, Australia and New Zealand are jointly developing a Spatial Data Infrastructure (SDI) under the auspices of Australia New Zealand Land Information Council (ANZLIC). The same principles have applied to the SDI as to other forms of infrastructure (e.g. roads and communications), namely that a single, consistent, accessible and government-funded infrastructure should provide the basis for competitive, private sector value-added services to be developed. ANZLIC has followed a decentralized approach in developing the Australian SDI with the aim of ensuring users requiring national coverage data will be able to access complete and consistent datasets from distributed databases under the custody of

Table 8.5 Selected digital geographical data relating to Australia and New Zealand

Company/Organization	Datasets description	WWW address
Remotely sensed		
Australian Centres for Remote Sensing (ACRES) part of Australia Surveying and Land Information Group (AuSLIG)	ERS (European Space Agency satellite) JERS (Japanese Earth Resources Satellite) LANDSAT, resolution 15 m, 30 m and 80 m depending on sensor RESURS Russian satellite, resolution 160 m SPOT-LITE (SPOT satellite imagery), resolution 10 m panchromatic and 20 m multi-spectral	http://auslig.gov.au/acres
Cartographic		
Australia Surveying and Land Information Group (AuSLIG)	Topographic mapping, 1:50 000; 1:100 000; 1:250 000 Elevation model, resolution 9 s of longitude and latitude (approximately 250 m); resolution 18 s of longitude and latitude (approximately 500 m) partial coverage; resolution 3 s of longitude and latitude (approximately 100 m) partial coverage. Contours, 20 m interval at 1:100 000	http://auslig.gov.au
Land Information New Zealand (LINZ) – National Topographic/Hydrographic Authority	Topographic mapping, 1:50 000. Land Information System Land*online*.	http://www.linz.govt.nz http://www.landonline.gov.nz
New Zealand Aerial Mapping Services Ltd.	Aerial photographs including scanned images.	http://www.nzam.com

Censuses and Surveys		
Australian Bureau of Statistics	Population Census	http://www.abs.gov.au
	Household Expenditure Survey	
	National Health Survey	
Statistics New Zealand	Census of Population and Dwellings	http://www.stats.govt.nz
	Agricultural Production Survey	
	Household Labour Force Survey	
	Household Economic Survey	
	International Visitors Survey	
	Time Use Survey	
Administrative and Statutory		
Statistics New Zealand	Building Consents Issued	http://www.stats.govt.nz
	Demographic Projections	
	Estimate of New Dwelling Units	
	Population Estimates	
	Vital Registration – Births and Deaths	

various jurisdictions. The SDI model has four key components – institutional framework, technical standards, fundamental datasets and clearing house networks. A fundamental dataset is one for which more than one government department or agency has a need for data with a consistent national coverage. These fundamental datasets are viewed as relating to a wide range of over 40 themes under a narrower series of topics including primary reference, administration, natural environment, socio-economic and built environment.

8.7 INDIA

The Government of India joined the 'space race' in the 1970s with the establishment of the Indian Space Research Organization (ISRO) with the principal objective of developing satellites, associated launch systems and ground receiving stations. Included within these series of satellites were some concerned with Earth and atmospheric observation namely the Indian Remote Sensing (IRS) satellite. The first in the series was launched in 1988 with subsequent missions in 1991, 1995 and 1997. In India the data from these systems are processed and disseminated by the National Remote Sensing Agency (NRSA) based near Hydrabad, which also handles data from foreign satellites and has aircraft with panchromatic cameras and multi-spectral scanners. Although there is a long history of ground survey for the purposes of topographic mapping, with the Survey of India having been established in 1851, and despite India's foresight in developing remote sensing satellites, it does not appear that the provision of digital data for mapping and GIS purposes has developed so extensively. Work is in progress on producing sheets in a national topographic map series at 1 : 25 000 scale which has been developed by dividing the 15′ × 15′ 1 : 50 000 maps into four blocks of 7′30″ × 7′30″. The Geological Survey of India has also mapped India's geology at a scale of 1 : 250 000 printed map sheets. A geoscience database has been developed covering almost the entire country at 1 : 50 000 scale; some sectors are available at 1 : 25 000.

The Population Census in India has been held on a decennial basis since 1872. The census authority operates a comprehensive internet site (see Table 8.6) that provides extensive information relating to the history of census taking in India and details, including the questionnaire schedule, of the most recent enumeration with a reference date of 0 : 00 h 1 March 2001. In addition there are mapped and tabular presentations of statistics down to district level, some of which relate to the latest census, although the majority are derived from the 1991 count. The government of India also operates a Sample Registration System (SRS) to provide an annual supplement of reliable birth, death, fertility and migration indicators to the decennial Population Census. The SRS was started on a trial basis in 1964/65 and then became operational in 1968/70. It has now grown to an annual sample of 6671 spatial units (2235 urban and 4436 rural) covering about 1.1 million households and 6 million individuals. The SRS data can be accessed via the CensusofIndia WWW site. The Population Census and Sample Registration System form only part of the Indian

Table 8.6 Selected digital geographical data relating to India

Company/Organization	Datasets description	WWW address
Remotely sensed		
National Remote Sensing Agency (NRSA)	IRS and foreign satellite data	http://www/nrsa.gov.in
Indian Space Research Organization (ISRO)		http://www.isro.org.in
Cartographic		
Survey of India	Paper maps 1 : 25 000 (partial coverage)	Website under construction
Geological Survey of India	Paper maps 1 : 25 000 (partial coverage)	http://www.gsi.gov.in
Censuses and Surveys		
Administrative and Statutory		
Department of Statistics	Population Census	http://www.censusindia.net
	Sample Registration System	
National Information Centre (NIC)	Economic statistics series	http://www.nic.in/stat

Table 8.7 Selected digital geographical data relating to selected other countries

	Company/Organization	Datasets description	WWW address
Remotely sensed Cartographic			
Europe	Multipurpose European Ground Related Information Network (MEGRIN)	Co-ordination role	http://www.megrin.org
Africa	Africa Data Dissemination Service (ADDS)	Maps data covering administrative boundaries, agro-climatic zones, crop land use, dem, hydrology, railroads, roads, vegetation. Normalized Difference Vegetation Index derived from satellite imagery.	http://www.edcintl.cr.usgs.gov/adds
Canada	Centre for Topographic Information at Sherbrooke (CTIS) of Natural Resources Canada	Toporama, updated road network, Canada3D, administrative boundaries, vector topographic maps	http://www.cits.rncan.gc.ca
The Philippines	Earth and Natural Resources Database (ENRDBAS)	Environment and Natural Resources Directory	http://www.psdn.org.ph/iemsd/dbas/enrnet2.html
Japan	Japan Oceanographic Data Centre (JODC)	Biology, geophysics and oceanography online data	http://www.jodc.jhd.go.jp
	Geographical Survey Institute	Elevation data, 50 m and 250 m resolution Administrative boundaries. 1 : 25 000; 1 : 250 000 Raster maps, 1 : 25 000 Topographic maps, 1 : 10 000, 1 : 25 000 and 1 : 50 000	http://www.gsi.gov.jp

The Netherlands	*Nationaal Clearinghouse Geo-Informatie* (NCGI)	Centralized clearing house providing online access to cartographic and statistical information	http://www.ncgi.nl
		Cadastral map, Registers and Topographic base map	http://www.kadaster.nl
Portugal	*Sistema Nacional de Informação Geográfica* (SNIG)	On-line access to thematic and topographic maps, and aerial photographs	http://www.sig.cnig.pt
Censuses and Surveys			
Administrative and Statutory			
Canada	Statistics Canada		http://www.statcan.ca
		Agricultural Census Population Census Also statistics on economy, land, people and the State	

Statistical Service, which combines central and devolved forms of administration. At the national level the Department of Statistics oversees the activities of bodies such as the Central Statistical Organization and the National Sample Survey Organization and co-ordinates the Indian Statistical Service. Much of the data collection is undertaken by the State Governments and their agencies. A Commission was set up in January 2000 to review the provision of statistical information under seven main headings, five of which are thematic and have the potential to involve the development of geographically specific information. In the case of agricultural statistics, it is recognized that more information for local areas is needed, for example in respect of more timely crop yield forecasting.

8.8 OTHER COUNTRIES

Countries in the previous sections were chosen to provide details of the extent of digital spatial and attribute data relating to geographical features and units in different parts of the world, and to indicate some of the trends occurring with respect to the availability of such information. Countries were not selected because they were considered to be particularly advanced in their approaches, but partly because they indicate the level of geographical detail in datasets that people are now able to access. Table 8.7 attempts to expand on the previous 'thumbnail' sketches of national digital spatial data coverage by providing summary information for more countries and hence indicates the broad range of digital data availability. The development of spatial data infrastructures, along the lines already discussed with respect to Australia, New Zealand, the UK and USA, is occurring in a number of countries and between countries, as in the case of Europe where the European Union has provided a stimulus to such cross-border actions. These developments in general have made finding out about the availability of spatial and geographical data that much easier and, as we shall see in the concluding chapter, are also having the beneficial effect of opening up access to data, in some instances free of charge, using the opportunities offered by the internet. With reference to Table 8.7, it is worth mentioning that the *Nationaal Clearinghouse Geo-Informatie* (NCGI) in the Netherlands is an example of how online access to geographical data is likely to progress in the future. It has an intuitive user interface, including an English version, although the mapped data and information that can be accessed have Dutch annotation. A wide range of topographic, cadastral and thematic information is available for the Netherlands from a single location.

PART 3
SECONDARY ANALYSIS

Chapter 9

Examples of Geographical Data Analysis

9.1 INTRODUCTION

The previous chapters have examined the characteristics of various types of geographical data and given some indication of their availability in a digital format in a selection countries. The aim of this chapter is to explore some of the problems that users of these data may encounter when undertaking GIS-based analysis. These problems can be divided into three main categories depending upon the type of geographical analysis being undertaken; these are referred to as comparative, temporal and integrative. These categories provide a structure for the following sections, which include a general review of the issues and an example of the type of analysis. Clearly there are as many potential applications using geographical data as there are users and the examples of these kinds of analysis have been kept relatively simple in order to emphasize the issues rather than to present complex and sophisticated investigations.

9.2 PROBLEMS OF COMPARATIVE GEOGRAPHICAL DATA ANALYSIS

Analysis of data relating to geographical features on the Earth's surface is often concerned within making comparisons between the same type of spatial unit. This might be as simple as looking at the differences and similarities of countries, cities, coastlines, oceans or any other category of geographical entity. The comparison may be concerned with examining the geospatial characteristics of the feature or the attributes associated with them. Figure 9.1 illustrates in general terms how line and

(a)

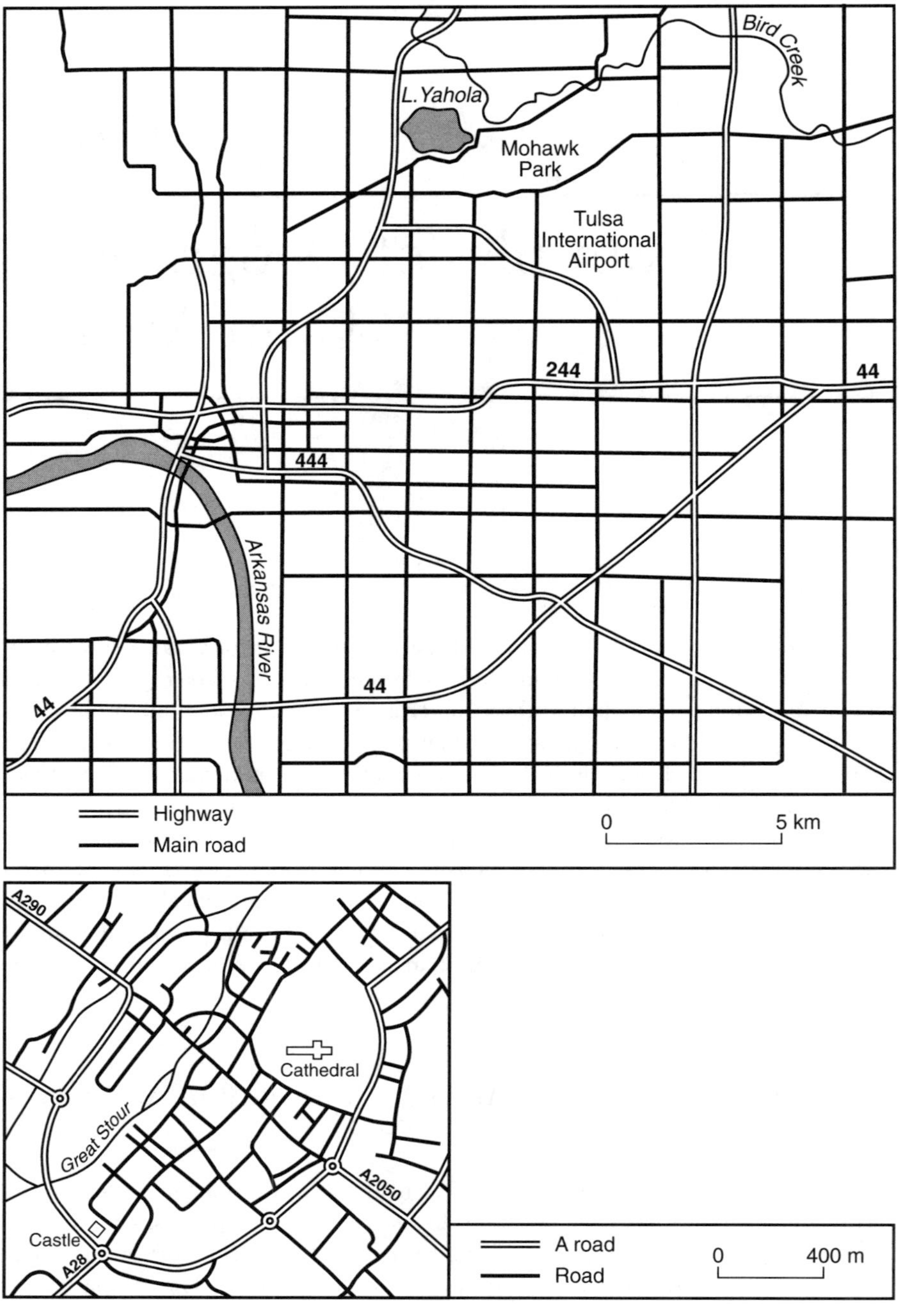

Figure 9.1 Variation in the physical form of geographical features within different cultural contexts. (a) 'Grid-iron' and medieval cities, (b) regular and irregular administrative units.

(b)

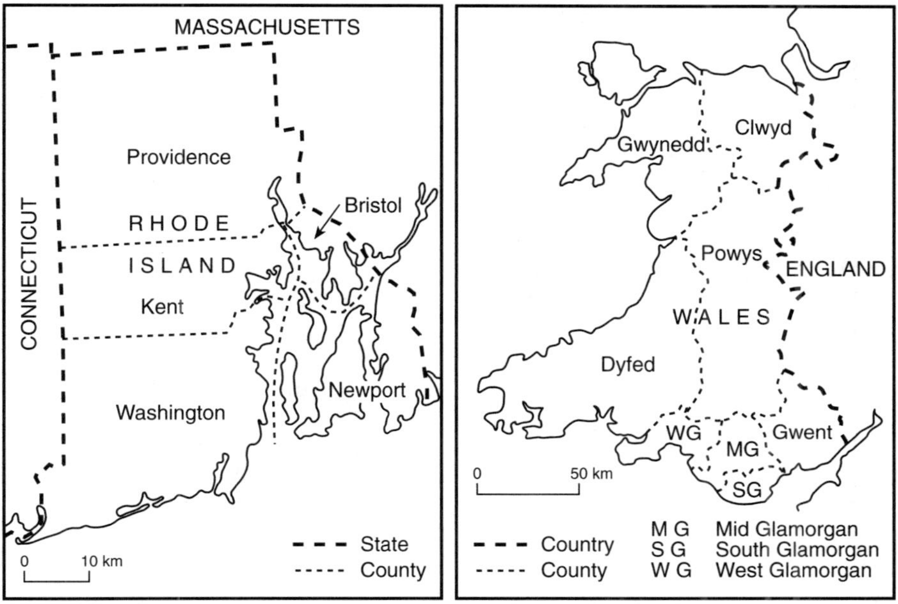

area features of the same class may differ across the Earth's surface at one point in time. The regular 'grid iron' pattern of streets in US cities is readily apparent (Figure 9.1a) and has been incorporated into city planning in a number of developing countries. However, this layout of streets contrasts markedly with cities whose cores reflect and perpetuate their medieval origins, such as Canterbury in the UK. Area features of the same or an analogous type may also display similar variation in form. Figure 9.1b illustrates this with respect to counties in the USA and the former counties in Wales. Although the administrative functions of counties in the two countries are culturally specific, they both serve as a type of administrative unit. In the USA many counties are bounded by straight lines, which change direction at right angles. These contrast with the generally irregular, sinuous boundaries of British counties that often follow physical features (e.g. a river or watershed) and may adhere to historical lines.

The spatial characteristics of geographical features can also vary within their extent at one point in time across relatively short distances. Figure 9.2a shows the road network in an area 15 km by 8 km (120 km^2) in Lincolnshire, England. The land to the east is extremely flat with spot heights of 2 or 3 m: here many of the roads are straight and change direction at right-angles, whereas the modest but relatively abrupt rise in altitude in the centre has produced a more irregular road layout and a series of village settlements. The straight road system resumes towards the west across a gently rising terrain. Similar local contrasts can be found for area features: Figure 9.2b illustrates that parishes in the area of the North and South Downs in southeast England frequently tend to have an elongated shape. This is a relic of the original

(a)

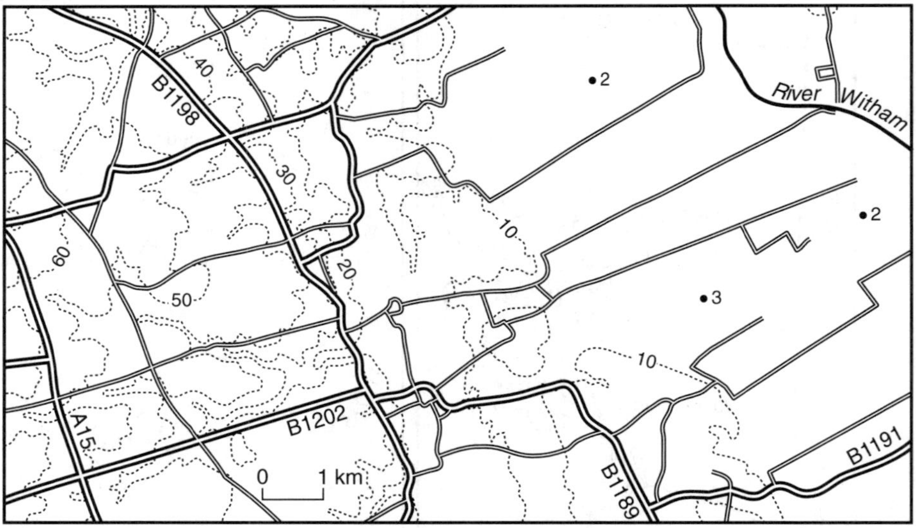

(b)

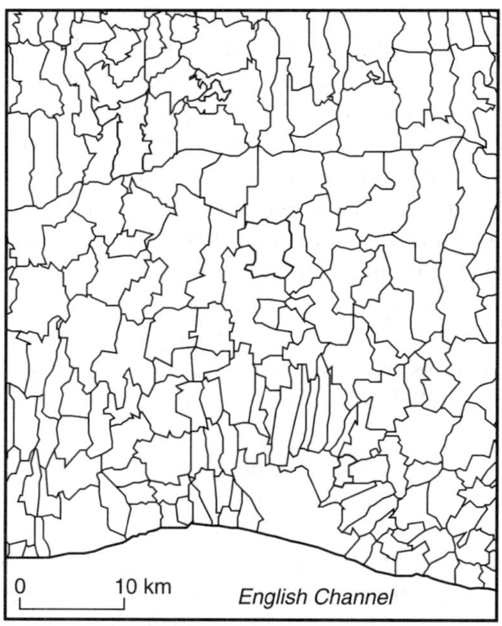

Figure 9.2 Variation in the physical form of geographical features within same cultural context over a short distance. (a) Road pattern, (b) parishes.

apportionment of land of differential altitude and fertility, and reflects the intention of sharing out good and poor quality land. In contrast parishes a few kilometres north in the central Weald part of southeast England tend to be more regular in shape.

Comparisons can be made between geographical units not only with respect to the same type of feature in widely separated and culturally distinct parts of the Earth as well as in close proximity to each other, but also with respect to the measurements of their attributes and variables. One of the basic characteristics of geographical description is its focus on comparison between different places in terms of the same attribute(s). We shall refer to such data interrogations as 'comparative geographical data analysis'. This type of analysis has been and remains one of the distinguishing features of geography throughout its development as a scientific discipline. At the current time, when the use of geographical data has become more widespread, we can see many different types of organization asking the basic questions:

- What are the characteristics of a particular place?
- How does one place differ from another?

Attempts to effect comparisons between different places and areas with respect to a range of themes are a regular, almost routine, activity of public and private sector organizations at the national and local scales. Within the literature the Canadian Geographic Information System (CGIS) is generally recognized as signalling the start of the GIS era (Tomlinson, 1988; Goodchild, 1995). Heywood *et al*'s (1998: 169) comment that the stimulus for the development of CGIS was 'the problem of measuring and *comparing* all this [mapped] information' (emphasis added) reveals the importance of comparison with respect to geographical data. Supra-national bodies, such as the European Union, are often exercised with making similar comparisons across state boundaries. One notable example from the relatively early stages of GIS development in Europe is the European Commission's Coordinated Information on the European Environment (CORINE) programme (Rhind *et al*, 1986; Mounsey, 1991). This project sought to create a general environmental GIS that would promote consistency and standardization in the collection of environmental information across the countries of the EC. As with CGIS, CORINE was initially concerned with creating an inventory of georeferenced information, but in this case the main concern was with nature conservation rather than resource capability assessment.

During the 1980s the European Commission also started to publish statistics obtained from its constituent countries and harmonized in a socio-economic database entitled *Regio,* which is based on the NUTS hierarchy of spatial units with the following structure when first started:

Level 1: Nation States (10)
Level 2: European Community Regions (64)
Level 3: Basic Administrative Units (167)
Level 4: Subdivisions of Basic Administrative Units (823)
Level 5: Individual towns or communities (97 812)

The statistics are organized in eight domains covering population, economic measures, employment and unemployment, industrial activity, agriculture, transport and participation in the Community's financial investment schemes (EUROSTAT, 1987). However, the difficulties of cross-boundary comparison are illustrated by the following extract from the notes to the 1986 publication relating to the seemingly straightforward concept of Total Population at NUTS Level 1:

> For the United Kingdom the data refer to the 'home population' as defined for the 1981 census; this concept differs from the total resident population . . . in one respect: 'home' population data include foreign armed forces stationed in the UK (EUROSTAT, 1986).

In total there were 39 notes of this type identifying where at least one of the 10 countries that comprised the EC at that time deviated from the 'norm', and all countries were referred in one note. As the European Union looks to expand its border eastwards to include states of the former Eastern Block, it seems likely that the difficulties of achieving consistency in *Regio* will be exacerbated.

Geographical comparison is about identifying the differences and similarities between places on the Earth's surface and determining whether or not these are significant. Such significance may be judged by means of statistical analysis techniques, or less formally by description. One of the fundamental problems is to ensure comparison of like with like. Following on from the previous discussion, one can argue that contrasts in the physical form of geographical features are a legitimate basis for comparison and may yield valuable insights. Thus, the different physical form of typical North American and European cities exemplified in Figure 9.1a is itself a basis for comparison, which can be achieved by means of the common spatial language of Cartesian geometry. However, one of the main problems arises with comparisons made regarding a particular theme or attribute, which may not easily be capable of definition and/or measurement in terms of shared procedures. On a superficial level users of geographical data across cultural boundaries need to be aware of differences of terminology, such as the terms duplex and semi-detached to identify joined pairs of houses in the USA and UK respectively. This type of difference is fairly easy to cope with, since it refers to conceptually the same type of housing in two cultural contexts, and provided that data users are made aware of the terminology then there is effectively no difficulty in reconciling the terms. However, more intractable problems of comparison can occur if there are fundamental differences of definition between countries, or even parts of one country. For example, a number of county and district local authorities or community councils in the UK have undertaken surveys of a range of facilities (shops, professional service providers, meeting places, etc.) during the last 40 years. Table 9.1 shows the extent of coverage in two surveys relating to Suffolk and West Glamorgan in 1978 and reveals some overlap in the range of facilities covered in each survey, but also significant differences. The Suffolk survey was primarily concerned with examining change and covered a limited range of facilities by means of a questionnaire sent to Parish Councils. The West Glamorgan example was published by the Welsh Consumer Council as part of an assessment of facilities across the whole of Wales and the information was collated

Table 9.1 Facilities included in Suffolk and West Glamorgan surveys in 1978

Suffolk	West Glamorgan
Education:	Education:
Primary School	Primary School
	Secondary School
Health:	Health:
General practice surgery	General practice surgery
	Pharmacy
	Dentist
	Optician
	General hospital (with/without accident and emergency)
	Professional Services:
	Chartered accountants
	Solicitors operating legal aid scheme
	Architects
	Chartered surveyors
	Estate agents
	Funeral directors
	Banks and building societies
	Public Offices and Services:
	Library
	Engineers and surveyors
	Social services
	Births, marriages and deaths Registry
	Planning
	Magistrates court
	Recreation and Entertainment:
	Adult education centre
	Cinema
	Community centre
	Golf course
	Indoor swimming pool
	Theatre
	Youth centre
Retail:	Retail:
General Store	Jewellers
Sub-Post Office	Florists
Public House	Shoe Repairs
Garage	Laundries/Dry Cleaners
Butcher	Hairdressers
Baker	Sub-Post Office
Mobile Stores	
	Public Transport:
	Bus service
	Rail service
	Welfare and voluntary services:
	Homes for the elderly
	'Meals on Wheels' kitchens
	Women's Royal Voluntary Service clubs
	Salvation Army

Source: Suffolk Rural Community Council; Welsh Consumer Council.

from a range of sources rather than a specific questionnaire. When examining the results of the surveys, it is not possible to tell whether a particular facility is lacking because it was omitted from the list covered by the survey or because it was genuinely not present. Similar problems can arise from the cultural specificity within which concepts have been defined. This problem is perhaps more common in respect of attributes associated with human activities or conditions, but is also possible in the case of environmental and physical phenomena. A spectrum of definitional differences in attributes can be envisaged between countries or regions: at one end there are attributes relating to geographically referenced entities about which there should be little scope for incompatibility or dispute, such as gender or soil particle size. At the other extreme, there may be inconsistency in the definition of the underlying concept. The 32 changes in the rules governing making a claim for unemployment benefit in the UK between October 1979 and October 1996 (see Table 9.2) illustrate how these definitional problems can arise within a single country, let alone across international borders.

There may be a relatively small or large amount of differentiation in how a particular concept is defined and operationalized for the purpose of data collection and reporting in different places. Hence the recent 2001 UK Census of Population defines a household as either a person living alone or a group of people with common housekeeping in terms of sharing at least one meal a day or a living/sitting room, whereas the US 2000 Census focused more on a person's usual place of residence and associated people living at an address with the concept of a household. These definitional differences may have a relatively minor impact on the reliability of geographical comparisons between countries, regions or other places on the Earth's surface. The problem is that users may not be able to determine the extent of the impact.

9.3 COMPARATIVE ANALYSIS PROJECT

This project is intended to illustrate some of the difficulties associated with trying to examine variations in particular geographical phenomena between different parts of the Earth's surface at one point in time. As we shall see in Section 9.4, what may be meant by the phrase 'at one point in time' is also problematic. However, for the present it is taken to refer to a fairly narrow temporal window of one or two years. The project concentrates on the twin problems of consistency in the definition of attributes and of geographical features and thereby aims to demonstrate that an initial, simple interpretation of what the information purports to show may be misleading and lead to invalid conclusions. Although the questions asked in this project are relatively straightforward, their seeming simplicity should act as a warning to those engaged in more complex comparisons between different parts of the Earth's surface of the pitfalls that exist for the unwary.

9.3.1 The Problem

The basic geographical units comprising local government areas in the UK were outlined when presenting the standard geographical hierarchy connected with the population census in the UK. In England there is essentially a twofold system with an upper tier comprising counties and a lower one of districts, which for historical reasons may sometimes be called boroughs. The two types of local authority are divided on a functional basis, in other words in terms of their responsibilities, although some activities, notably planning, are split along strategic and implementation lines between the two. This system was established across England and Wales in April 1974 and in 1975 in Scotland, where the upper tier units had larger surface areas and were referred to as regions. In 1986, the six Metropolitan Counties and Greater London were abolished and were replaced by a single tier of administration comprising the Metropolitan Districts and London Boroughs. In the mid-1990s further changes were made with comparable single tier Unitary Authorities introduced as separate enclaves within some counties, for example Southend-on-Sea became separate and independent of the county of Essex. Some English counties ceased to exist as a result of these changes, such as Berkshire, which persists only as a titular county, and they disappeared altogether in Wales, as did regions in Scotland. Further change occurred in 2000, when a Greater London Authority was created with a strategic role across the whole of the metropolitan area covered by the former Greater London Council.

This project concentrates on one tier in this complex system, namely the districts within the rural shire counties, where for the most part the two-level hierarchy persists. The analysis is further constrained by concentrating on part of a single region. This project explores the problems arising from the fact that, although these districts are functionally equivalent across the country, they vary in character in respect of their land use pattern and their socio-economic composition. The particular problem that can arise when comparing statistics for these units relates to the fact that some may comprise a mixture of rural land uses with either a scattering of relatively small towns or one or two large towns, whereas others may relate to a moderately large provincial city or town. In the latter case it is quite possible for the boundary of the local authority to cut through the outer fringes of the settlements' built-up area, with these portions assigned to one or more surrounding authorities that may be largely rural in character. Figure 9.3 covers the part of the Eastern Region formerly referred to as East Anglia and comprising 20 districts within the counties of Cambridgeshire, Norfolk and Suffolk. The three districts of Cambridge, Ipswich and Norwich are clearly identifiable as small in surface area in relation to others within the study area.

9.3.2 Data Sources

One solution to the problem of incompatibility between the boundary of a local authority district and the physical extent of its built-up area is to define a completely

Table 9.2 Definitional changes affecting the level of unemployment claimants October 1979 to October 1996

Date	Change	Estimated effect Increase/decrease
October 1979	Benefits paid fortnightly	+20 000
October 1979	Seasonal total adjustment	−20 000
February 1981	Removal of people in training or temporary work	−370 000
July–October 1981	Seasonal figures adjusted to allow for DHSS strike	−20 000
July 1981	Males 60+ claiming supplementary benefit given option of not signing on	−30 000
July 1982	Unemployment benefit taxed, single parents encouraged to opt for untaxed supplementary benefit	NE
October 1982	Computer count introduced	−190 000
October 1982	Seekers of part-time jobs removed from count	NE
April 1983	Males 60+ not entitled to unemployment benefit no longer required to register for NI credits	−107 000
June 1983	All males 60+ allowed long-term supplementary benefit	−54 400
June 1983	School-leavers able to claim in September after leaving school	−200 000
October 1984	Community programme entry limited to people on unemployment benefit	−29 000
July 1985	Northern Ireland records reconciled with DHSS system	−5000
July 1985	Unemployment benefit paid in arrears	NE
March 1986	Statistics delayed two weeks	−50 000
June 1986	New method of calculation (larger denominator)	−1.4%
June–October 1986	Tests tightened for Restart and work availability	−300 000
October 1986	People with insufficient NI credits not entitled to reduced unemployment benefit	−30 000
October 1986	Voluntary unemployment disqualification extend to 13 weeks	−9000
April 1988	Voluntary unemployment disqualification extend to 26 weeks	−12 000
April 1988	Part-time work category reduced to 24 hours per week	NE
June 1988	New method of calculation (larger denominator)	−0.1%
September 1988	16- and 17-year-olds barred from claiming unemployment benefit	−120 000
October 1988	Conditions for short-term benefits changed	−38 000
October 1988	Some persons 55+ put on pensions rather than benefit	−30 000

Table 9.2 (*cont.*)

Date	Change	Estimated effect Increase/decrease
September 1989	Proof required of actively seeking work	−25 000
September 1989	Work availability tests tightened for repeat claimant	−350
September 1989	Job offer refusal not permitted on grounds of low wages	0−25 000
September 1989	Redundant Miners Scheme claimants not required to sign on	0−28 000
September 1990	Unemployment benefit and income support ended for students	−100 000 per annum
April 1996	Unemployment benefit period reduced to 6 months	−70 000
October 1996	18–24-year-olds benefit reduced by introduction of Jobseeker's Allowance	−180 000

Notes: NE denotes no estimate; estimated effects relate to time of introduction.

Figure 9.3 East Anglia study area for comparative analysis project.

new set of spatial units based on the functional economic and social connections between metropolitan, urban or provincial centres and their hinterlands. Such an exercise was undertaken in the 1980s using census statistics for small spatial units by the Centre for Urban and Regional Development Studies at Newcastle University (Coombes *et al*, 1982). Although this functional regionalization was an important exercise and these units have been used as aggregation areas in some statistical series, their application has not been widespread apart in respect of employment. This project presents an alternative approach that involves linking together digital boundary files for the local authority districts and for built areas within the region. The physical extent of built-up areas was defined by the British census authorities following the 1991 enumeration in order to publish key statistics for so-called urban and rural areas. The boundaries of these built-up areas have been digitized and the upper part of Figure 9.4 illustrates that these are scattered and relatively small within the study area, although a few larger urban settlements are evident. Comparing the extent of built-up land in the districts of Cambridge, Ipswich and Norwich in the lower part of Figure 9.4 reveals that in the first instance the district boundary almost entirely contained the urban area, whereas there were some sizeable overlaps of the Ipswich and Norwich boundaries. In addition there were also a few instances where other settlements straddled district or county boundaries.

A substantial number of statistical sources include information at the local authority level and hence the implications of these mismatches between district boundaries and the extent of built-up areas could be considered in respect of several themes. The statistics used in this project are the number of planning applications considered by the local authorities in 1998 as assembled and published by central government (DETR, 2000a), which were considered as administrative data in Chapter 7. These statistics are not normally published for areas smaller than local planning authorities and hence it would be difficult to estimate figures for relatively large urban settlements that extend beyond their district boundaries other than by some form of GIS-based manipulation. It might seem reasonable to assume that the majority of planning applications would relate to existing built-up areas and to places fringing these zones. However, the aim of this project is simply to produce figures relating to 'complete' towns and cities that overlap administrative boundaries.

9.3.3 Analysis and Results

The basis of the spatial data manipulation carried out in this project is that by overlaying and intersecting two (or more) different polygon networks it is possible to estimate the values of attributes or variables for the newly created spatial units. This type of analysis represents one of the basic spatial data analysis operations and can be used in a number of different situations. In this case, the objective has been to estimate the number of planning applications associated with those parts of cities and towns that are 'detached' from the rest of their contiguous built-up area by an administrative boundary. Table 9.3 summarizes the number of applications per local planning authority as reported in the published tables and as estimated by the overlaying procedure by means of area apportionment. The differences between the two sets of figures are relatively minor since the areas of overlap are small. Ipswich and Norwich show the most significant adjustments, although even these are under 1.0%, where some planning applications have been reassigned from neighbouring districts to the contiguous built-up area of these settlements.

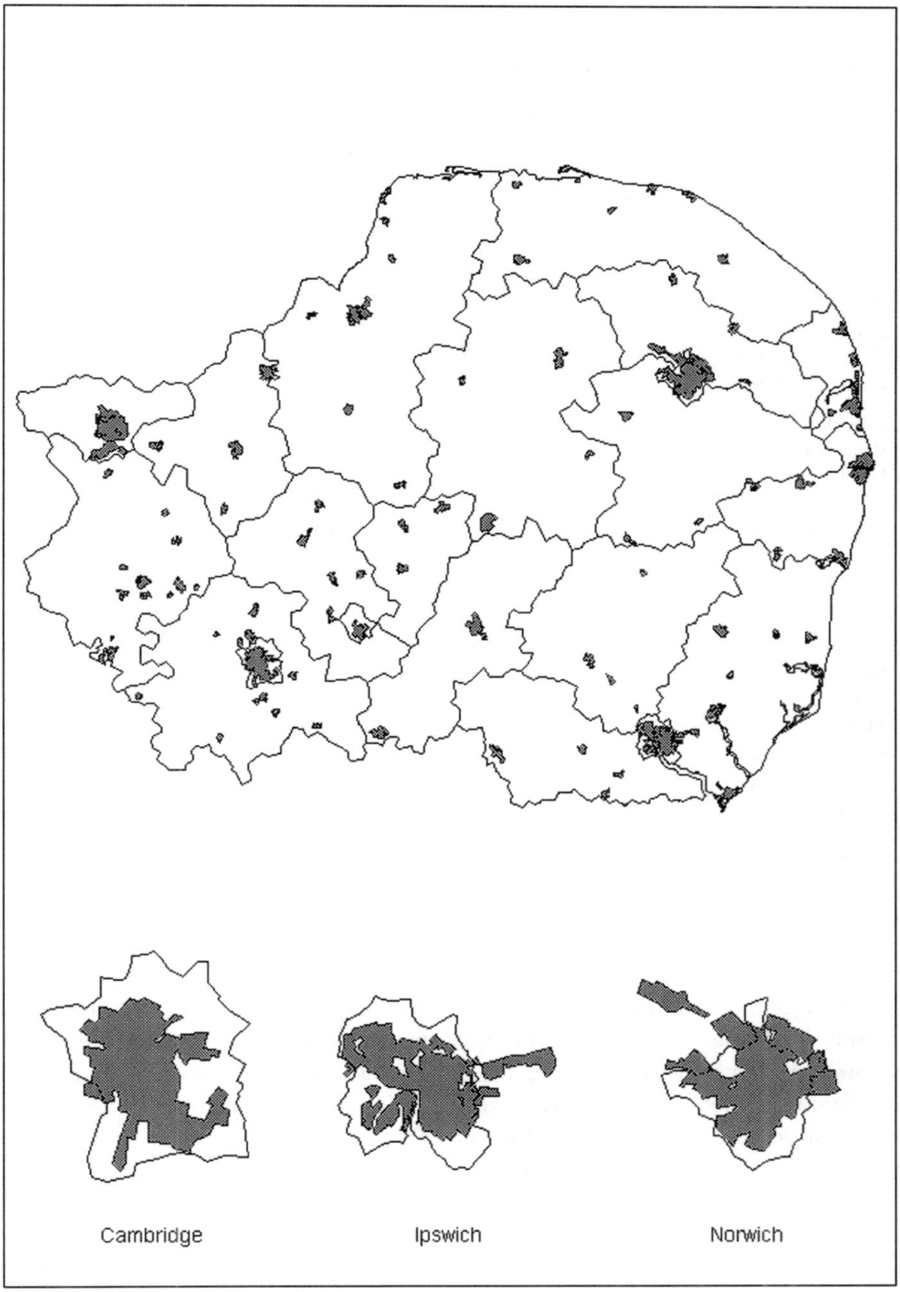

Figure 9.4 Mismatch between local authority boundaries and built-up areas.

Table 9.3 Comparison of number of planning applications considered in East Anglia district local authorities, 1998

	Planning applications 1998	Adjusted planning applications 1998	Difference	%
Cambridge	1109	1113		0.36
East Cambridgeshire	1061	1061	0	0.00
Fenland	956	957		0.10
Huntingdonshire	1770	1768	−2	−0.11
Peterborough	2090	2090	0	0.00
South Cambridgeshire	1328	1325	−3	−0.23
Cambridgeshire	8314	8314	0	0.00
Breckland	1468	1468	0	0.00
Broadland	1392	1366	−26	−1.87
Great Yarmouth	881	881	0	0.00
King's Lynn and West Norfolk	1549	1548	−1	−0.06
North Norfolk	1536	1537	1	0.07
Norwich	906	934	28	3.09
South Norfolk	1706	1702	−4	−0.23
Norfolk − 33 + 30	9438	9436	−2	−0.02
Babergh	1406	1406	0	0.00
Forest Heath	596	596	0	0.00
Ipswich	898	906	8	0.89
Mid Suffolk	1646	1645	−1	−0.06
St Edmundsbury	1318	1318	0	0.00
Suffolk Coastal	1715	1708	−7	−0.41
Waveney	1097	1099	2	0.18
Suffolk	8676	8678	2	0.02

This analysis has indicated how a misleading impression of the extent of development in Ipswich and Norwich might be obtained if only the published figures for these planning authorities were examined. There is less of a potential problem in the case of Cambridge where the large majority of the built-up area lies within the district boundary. The approach outlined here potentially has a wider application to other situations in which spatial analysis can help to provide an alternative perspective. Such administrative and electoral boundaries are, of course, periodically re-drawn under the authority of the Local Government Commission for England and equivalent bodies in other parts of the UK. Periodic electoral reviews are intended to ensure reasonable comparability in the number of electors represented by each councillor elected to a council within and between local authorities. A structural review of local government may be directed by the Minister of State, as occurred between 1992 and 1995, and has the aim of determining whether or not a single, unitary authority would provide a better reflection of community identities and interests rather than two councils.

9.4 PROBLEMS OF TEMPORAL GEOGRAPHICAL DATA ANALYSIS

One of the key characteristics of data relating to the Earth's geographical features is that these can change over time in two fundamentally different ways. Geographical features may be thought of as 'populating' the Earth's surface, in the sense that they occupy some bounded area of physical space. This notion relates to the classical view of space as absolute, comprising a 'container' in which objects or entities occur. Harvey (1969: 208) attributed 'much of the philosophy of geography . . . [to] . . . a "container" view of space . . . associated with the concepts of Newton and Kant'. According to this view, it is the contents of the spatial container that change over time and their position can be determined by means of the Euclidean geometry of a Cartesian referencing system or grid. Thus a particular portion of the Earth's surface may be occupied by a collection of co-existing but not necessarily connected entities. In this case the question of change over time is not so much concerned with changes in the relative location of the objects themselves, but with whether they are present or absent.

These ideas gave rise to the classical regional geographies of the 19th and 20th centuries in which the main focus was on identifying the assemblages of geographical phenomena occurring in different and diverse regions of the Earth. Many examples can be found in the literature of where the regional concept has been employed in geography. Butland (1960: 243) illustrates the genre with respect to Argentina when stating that the 'climatic contrasts and the variations in economic development dependent on historical evolution and trends differentiate the country into at least four well-marked geographical regions'. Descriptions of these areas combine information about their physical characteristics and their human population. The association between three of Argentina's regions, including their sub-divisions and relief units, are shown in Figure 9.5. It is notable that region 3, comprising an area mainly under 200 m, was regarded as a pastoral zone with its sub-regions defined in terms of different types of arable crops. According to the regional approach environmental characteristics or attributes, such as climate, soils and vegetation, were used to define or create a regional partition of the Earth's surface by means of classification. These regions took on the existence of geographical features and could be treated as land units.

The second type of change relates to the precise location, physical form and shape of geographical features (Raper and Livingstone, 1995; Massey, 1999). The feature continues to exist at different time periods, but its position relative to the georeferencing system changes. This can have ramifications with respect to the larger 'geography' of which the feature forms a component and recognizes that space is a relative concept not dependent upon a fixed spatio-temporal referencing system. Individual features may thus have changing relationships with each other according to how the 'distance' between them is measured. This notion argues that physical distance, measured in terms of metres and centimetres (or feet and inches), is but one of many 'metrics' that may be used. Hence, features may be separated from each other in

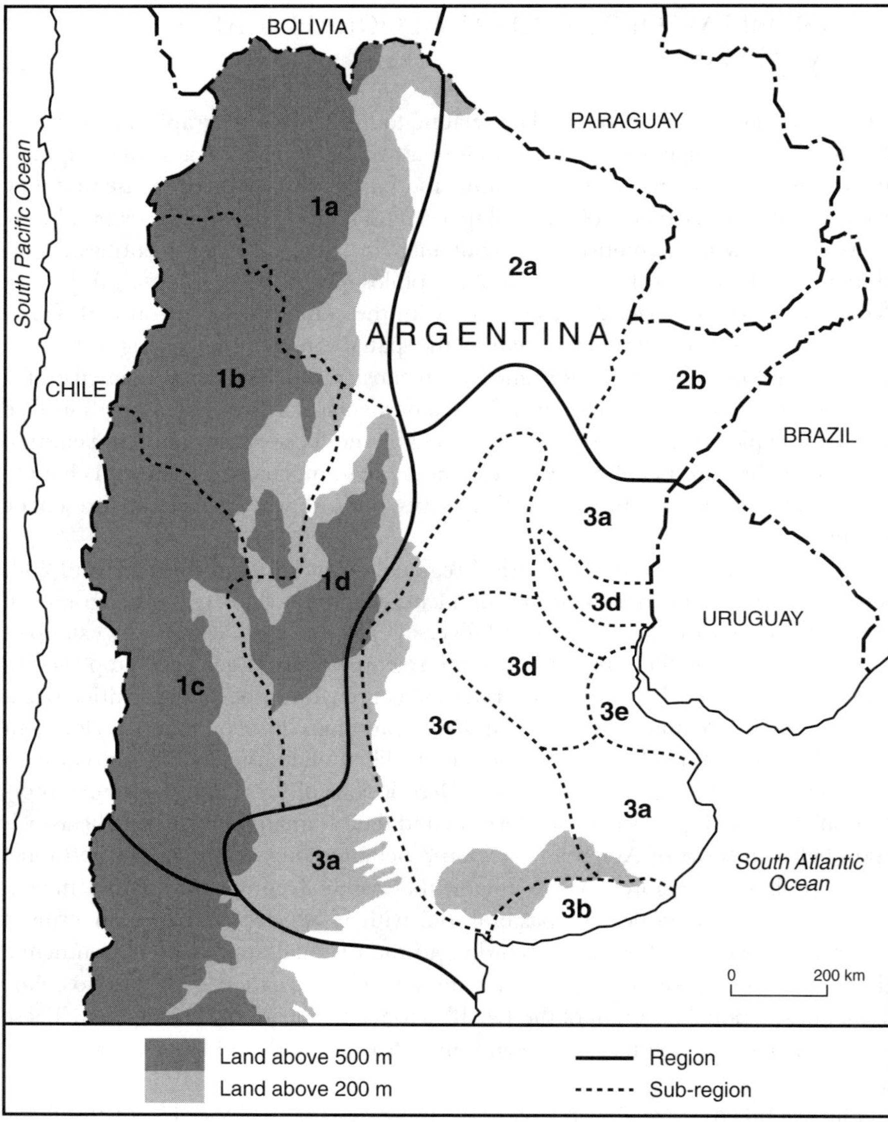

Figure 9.5 Classicial geographical regions.

terms of the time taken to travel between them or the financial cost of such movements. Figure 9.6 illustrates this generic problem by looking at how individual geographical entities of different types may persist but change their physical form and location over time, and how entirely different entities may occupy the same area of geographical space at different time periods. Springs emerging at the boundary between permeable and impermeable rocks illustrate how point features can also change their location and be present or absent at different times (Figure 9.6a). In wetter seasons, when the water table rises, springs at higher altitudes will flow, whereas these may stop flowing when the water table falls and new springs may emerge at low levels. Prolonged extraction of water from aquifers for human consumption may result in the long term in permanent extinction of springs. In some of the world larger water bodies are also susceptible to such changes. Figure 9.6b illustrates the migration of the Aral Sea in relation to the graticule: the main water body has decreased and moved eastwards over the last 50 years and an extensive salt flat has developed. Features created by human design can also evolve in a similar way. The route of a road may be altered temporarily by a diversion on account of a short-term obstruction or permanently by re-construction to by-pass towns, as illustrated by the course of the A27 trunk road being shifted by the construction of a by-pass to the north of Brighton (Figure 9.6c). The physical route of the road formerly known as the A27 running through the built-up area of Brighton and Hove was renamed as the A270 in the 1990s and a new route for the A27 constructed to the north. The final example, Figure 9.6d, illustrates how an administrative unit, in this case a county, can change over time. Before 1963 the counties of Essex, Kent, Middlesex and Surrey extended into the built-up area of London, then the Greater London Council (GLC) came into existence with a boundary extending outwards and absorbing parts of these counties. The GLC was abolished in 1986 and its functions were transferred to 33 London Boroughs that had been part of the lower tier of local government during the period 1974–86.

The way that changes in the physical form and incidence of geographical phenomena have been depicted in Figure 9.6 implies that at some point in time they can be regarded as fixed and incontrovertible. However, in much the same way that the relative view of space has superseded the absolute, so a relative conceptualization of time is now regarded as more useful than an absolute one. In the past it has generally been sufficient to 'date stamp' geographical datasets to a particular year, month, day or possibly hour and to accept that temporal variation in the measurements relating to the phenomena within these absolute units of time are insignificant. However, the diurnal rotation of the Earth around its axis means that the temporal reference of a particular location is constantly changing. For practical purposes the Earth has been divided into a series of time zones (Figure 9.7) whose boundaries mainly follow a single meridian over the oceans, although there is some deviation around parts of continents. Some of the larger landmasses, such as North America and Eurasia, span a number of time zones. For example, the USA is covered by five time zones with the east coast 4 hours ahead of Alaska and 5 ahead of Hawaii. It is apparent from Figure 9.7 that the time zone boundaries deviate from the meridians over many of the landmasses. The case of Iceland illustrates how that country's historical connections with Europe led to its assignment to the same time zone as western and central European countries despite its eastern coast lying some 14° west of the Greenwich (0°) meridian. The principle underlying the existence of these time zones is that places on the Earth should be in daylight when people are awake and in darkness when they are asleep. However, shifting patterns of daily activity amongst sections of the human

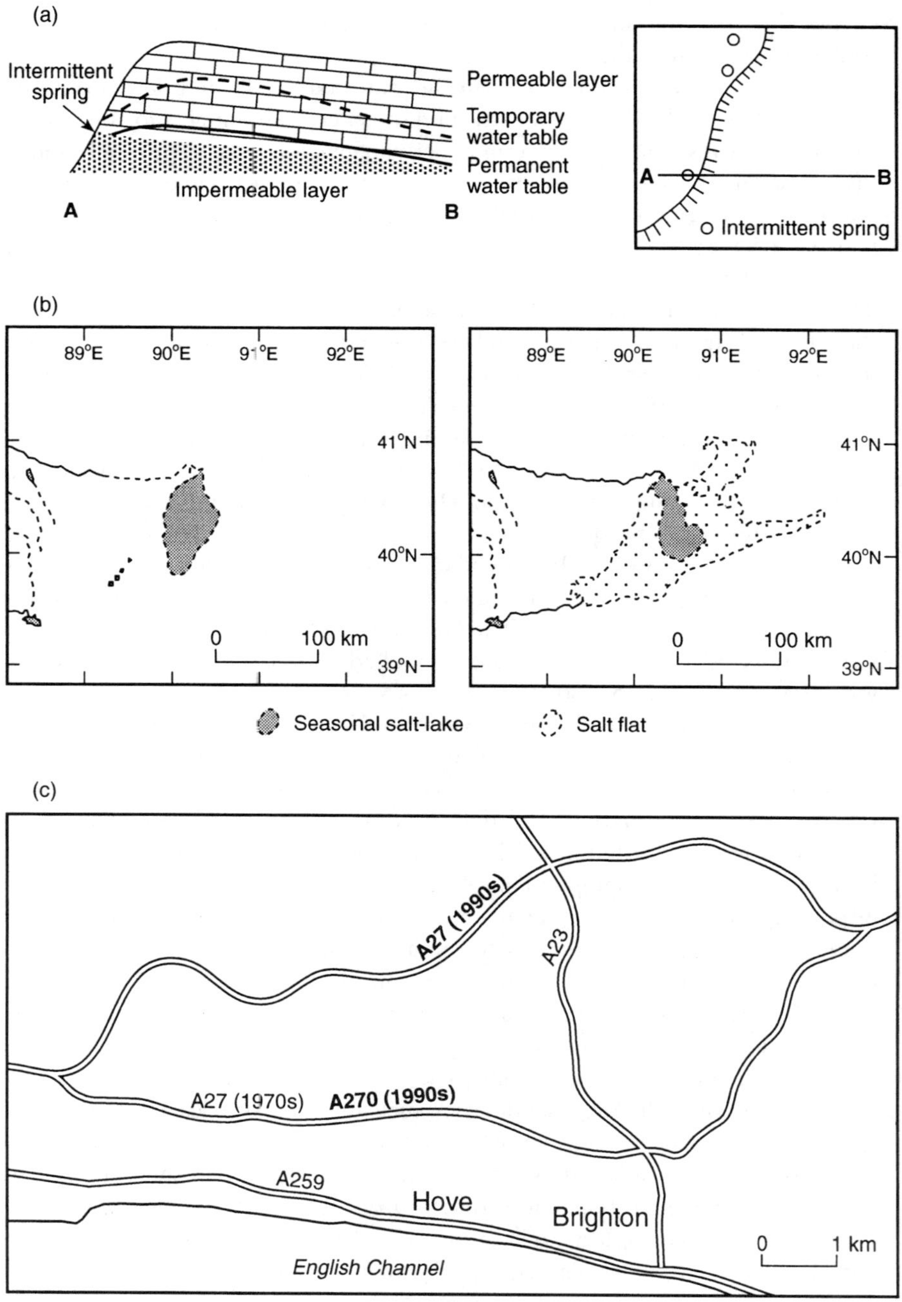

Figure 9.6 Change in the geographical entities over time. (a) Ephemeral springs, (b) water bodies (Aral Sea), (c) major (trunk) road, (d) administrative county.

(d)

1 Barking and Dagenham
2 Camden
3 City of London
4 City of Westminster
5 Greenwich
6 Hackney
7 Hammersmith and Fulham
8 Haringey
9 Islington
10 Kensington and Chelsea
11 Kingston upon Thames
12 Lambeth
13 Lewisham
14 Newham
15 Redbridge
16 Richmond upon Thames
17 Southwark
18 Tower Hamlets
19 Waltham Forest
20 Wandsworth

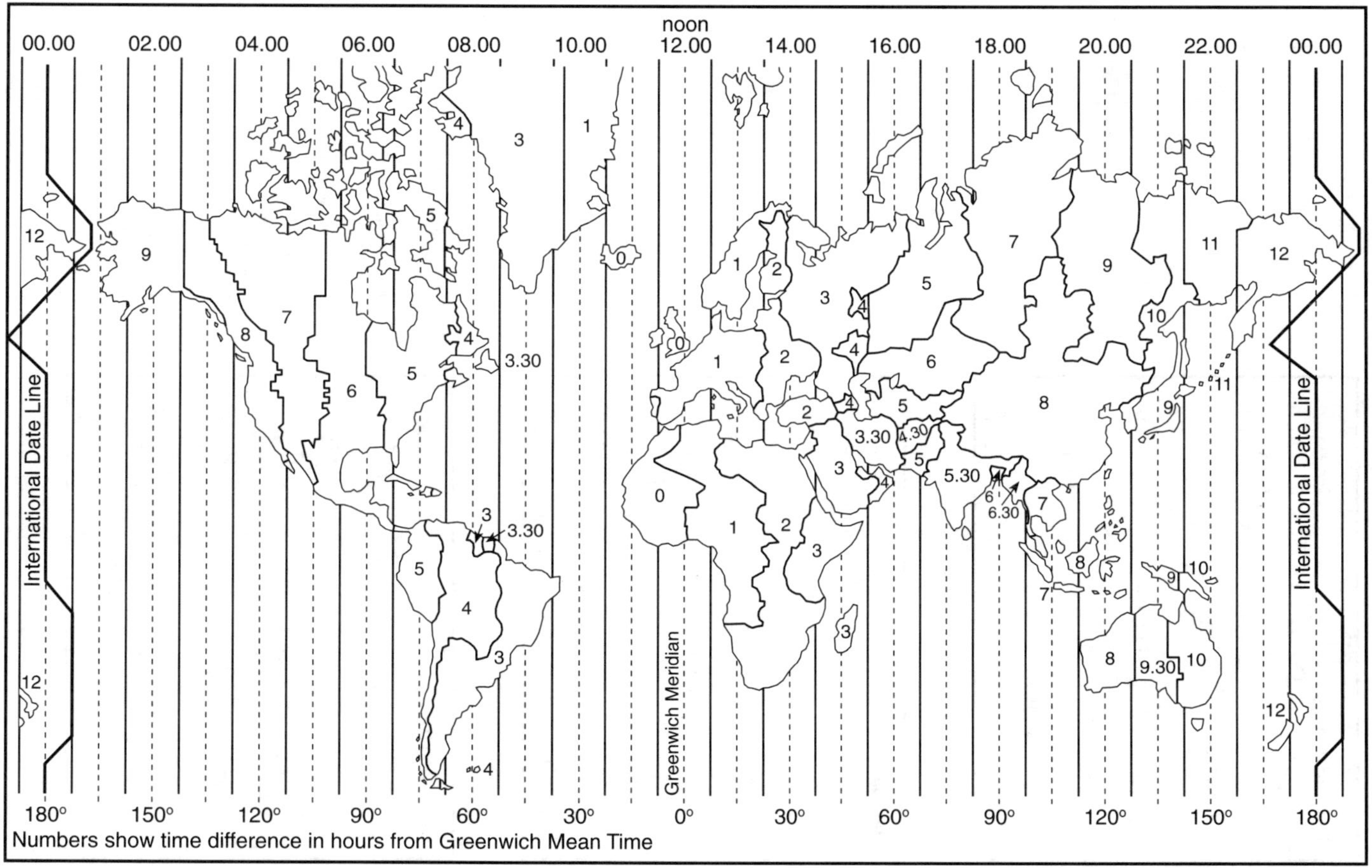

Figure 9.7 International time zones.

population are leading to the emergence of a 24 hour society and producing a situation in which some people spend the hours of darkness in work and of daylight asleep. Irrespective of the economic, social and psychological implications of these adjustments, it is likely that precise temporal references will become more important in georeferenced data.

Variations can occur in respect of a particular attribute or theme for the same geographical area over time. In other words geographical features can change with respect to their measured attributes or variables, even if these have been defined consistently. Hence, changes in the relative locations of the Aral Sea and A27 road depicted in Figure 9.6 may be further complicated by differences in the values of attributes and variables. For example, the number of vehicles on the road passing given points may have increased or decreased independently of the feature's re-location in geographic space. In some cases a change in these attribute values may be associated with an adjustment to the physical dimensions of the feature. An increase in the population size of cities and towns is likely to be linked to a concomitant expansion of their built-up area to cope with the additional people, by the provision of housing, workplaces and other types of infrastructure. Such adjustments may occur solely with respect to the internal structure of the entity, hence an increase in population may involve the construction of additional housing, but result in no alteration of the settlement's overall area. Equally there may be an overall enlargement with or without internal adjustment. However, change in some attributes do not alter the physical form or location of the geographical entity, but simply result in different values being recorded. These changes can occur with respect to attributes and variables on the four scales of measurement, thus the renaming of Leningrad to St Petersburg represents a change on the nominal scale, whereas the difference between the population of Britain in 1981 and 1991 is an increase on the ratio scale.

9.5 TEMPORAL ANALYSIS PROJECT

This project is designed to illustrate the problems of changes in the location of geographical features over time. It concentrates on the issues associated with matching between geo-spatial entities within a particular study area in order to enable analysis of statistical information. The project addresses the 'feature matching' problem with reference to spatial units that have been defined for administrative purposes. Previous chapters have shown that areas are frequently used as statistical reporting units, especially in the case of data connected with administrative, economic, political and social processes, and so the discussion may be extended from the specific to the general. Further, the principles that are exemplified by this project are equally applicable to situations involving point or linear features, and to spatial units associated with the natural environment as well as those arising from human activity.

9.5.1 The Problem

The purpose of this project is to investigate patterns of demographic change for small areas between the 1971, 1981 and 1991 Population Censuses in the UK. The project is designed to illustrate the problems of constructing a set of comparable geographical areas over time. The project concentrates on a particular region of the country, namely a rural area in mid-Wales, although the procedures and techniques could be extended and adapted to cover a larger area or be undertaken with respect to other types of geographical attribute data. Concentration on a particular area of the UK not only allows for a clearer exposition of the procedures involved in 'feature matching', but also avoids the difficulties associated with the fact that constituent countries of the UK have experienced different changes in their boundary systems and that census geography has been developed in separate ways. The project demonstrates the problems that arise when the boundaries of different types of polygon system change over time. In 1971 there were over 1500 local authorities in Great Britain divided between counties and smaller units known as either County Boroughs, Municipal Boroughs, Urban Districts or Rural Districts. Figure 9.8 shows these administrative units in the study area for 1971 and the smaller parishes/communities in 1981 and 1991 (LAs with solid lines, parishes/communities with dashed lines). Rees (1998: 1786) suggested that examining intercensal population change is 'one of the most important functions of a census'. As we have seen elsewhere in this volume, the main problems relating to the analysis of change concern discontinuities in the delimitation of geographical entities, the definition of population bases and variables, and thematic content.

9.5.2 Data Sources

In the interest of avoiding complexity that might cloud the main issues, only relatively simple population counts from the 1971, 1981 and 1991 UK Population Censuses have been used in the present analysis, although even these suffer from changes in the population base (see Section 6.2.1). Across the study area as a whole there was a −2.9% difference in the 1981 Census between the total population according to the 1971 persons present base and the 1981 usual residence base. In the 1991 Census there was a difference of −3.1% between the 1971 persons present base and the 1981 usual residence base, whereas the 1991 usually resident base, by taking account of absent households, produced a total population count that was 2.8% higher than the 1981 equivalent. These figures suggest it is reasonable to assume that population changes in excess of ±3.0% are not an artefact of the data. The population counts used in this analysis represent usual residents in 1971, 1981 and 1991, the latter according to the revised usually resident population base.

Innovative approaches to defining output areas for the 2001 UK Population Census and of linking back to 1991 were also examined in Section 6.2.1, but analysis further back across the decades is hampered by the lack of address georeferencing. It therefore becomes necessary to connect geographical areas over time and to estimate

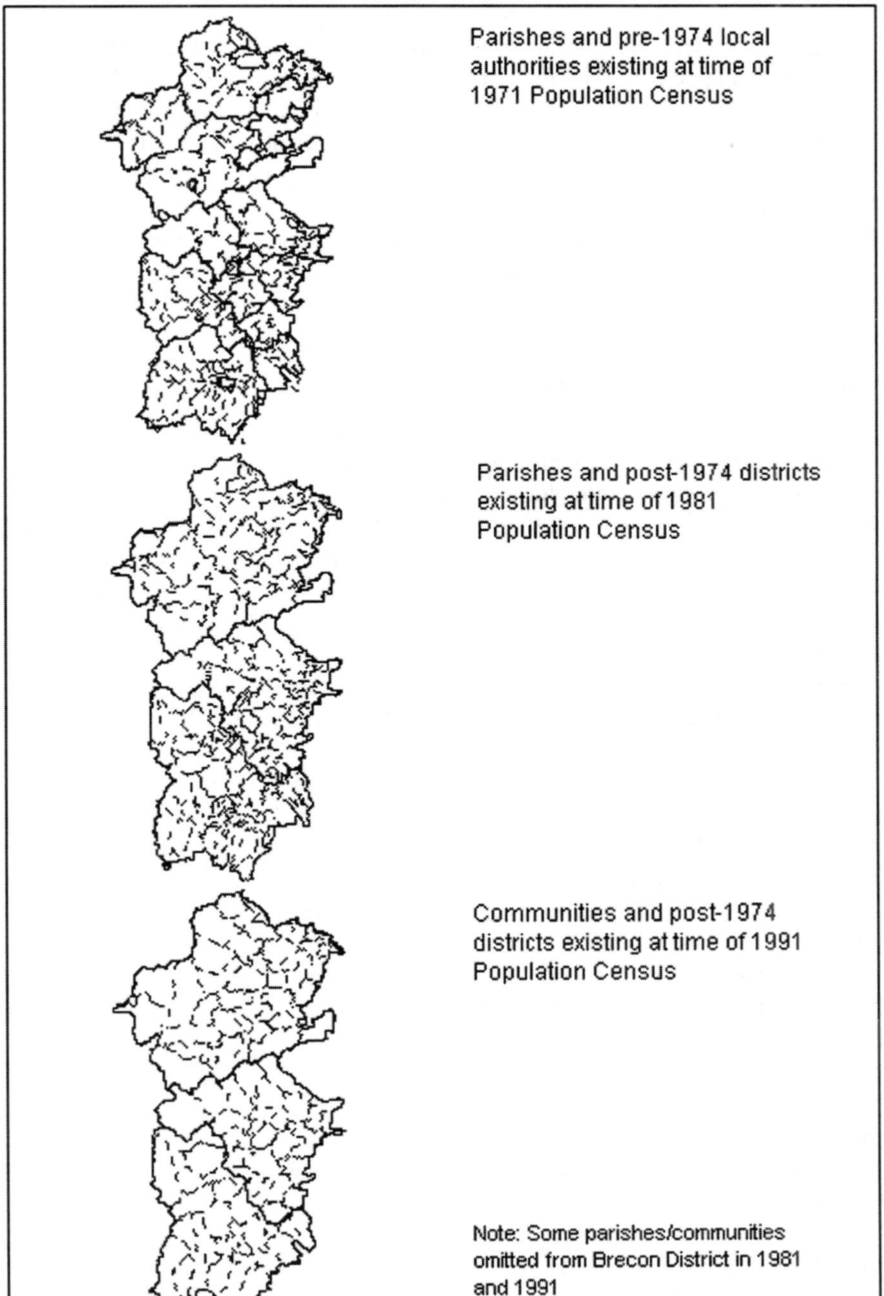

Figure 9.8 Study area for small area population change project.

population counts. Therefore the second important type of data used in this project is the digital boundary file for the small geographic entities in the study area. In particular the digital boundary files for the 1981 Census wards and the 1991 Census EDs have been accessed through the Census Dissemination Units at the Universities of Manchester and Edinburgh, and a digital file of the 1981 EDs created by Powys County Council was also obtained. Computer mapping was in its infancy in 1971 and hence only paper maps of the detailed census geography of the time were produced. However, an important means of connecting between the geographies of 1971 and 1981 Census in England and Wales are the Census Tracts, which were defined in the 1981 Area Master File in terms of EDs.

9.5.3 Analysis and Results

The focus of this analysis is on reconstructing the small area geography of the study area between 1971 and 1991 with a view to examining basic population change. The smallest geographical unit used for reporting statistics from the 1971, 1981 and 1991 Censuses was the ED. Although there is some continuity in EDs between successive censuses, they are not in themselves suitable for the present analysis because nationally only 46% of EDs were comparable from 1971 to 1981, and 40% from 1981 to 1991. Although it would be feasible to digitize the 1971 EDs from paper maps, the definition of the 1971–1981 Census Tracts in terms of EDs from each of these censuses and the existence of an alternative and reasonably consistent small geographical unit, the parish or community in Wales, opens the opportunity for an alternative strategy. Parishes (communities) would also be a suitable unit for extending the analysis further back to the 1961 or 1951 Censuses when ED statistics were not published nationally. Census Tracts were defined as the smallest comparable or identical areas linking between the 1971 and 1981 Censuses in England and Wales and were composed of either a single ED in both years or groups of EDs from one or both censuses. In rural areas the tracts were made equivalent to parishes (communities) in other areas they could be amalgamated to the pre-1974 local authorities (Municipal Boroughs (MBs) and Urban Districts (UDs)). In the study area 58.5% of the tracts comprised a single ED in both censuses and a further 18.6% were either a single ED in 1971 and two in 1981, or *vice versa*, or two EDs in both. These connections enable a set of comparable parishes and other 'old' local authority areas to be defined for the study area. In the 1980s there was a major revision of parish (community) boundaries that resulted in a reduction from 191 in 1981 to 100 in 1991 and brought about a 114.9% increase in mean population and 90.5% in area. The re-drawing of community boundaries presents the problem of whether to reconstruct the small area geography of the study area for the former, smaller communities, which had an average of 491 persons and an area of nearly 25 km^2 in 1981, or to collapse these into the newer, larger set. The latter approach has been followed to provide some consistency with contemporary data sources for the larger areas.

The essence of the analysis involves using such spatial analytic tools as point-in-polygon searching and polygon overlay to associate, on the one hand, centroid grid references or, on the other, the digital boundaries of one set of polygons with another. Initially it might seem that the generation of population counts for 1971, 1981 and 1991 Censuses aggregated according to the 1991 communities simply involves using the polygon overlay operation to assign the figures from the different censuses to the desired units by means of area apportionment. However, the study area is a generally mountainous region with a low overall population density (13 persons per km^2) and a 'very dispersed settlement pattern' (Williams, 1985: 127) with a large majority of people living in small settlements interspersed with isolated farmsteads and occasional clusters of housing. Simple re-aggregation through proportional assignment by means of polygon overlay could easily inflate the population count of a community because of its size, when in reality a substantial part of its area was uninhabited. This analysis therefore combined point-in-polygon searching of population weighted centroids with polygon overlay in order to generate the population counts.

A series of geocomputational procedures has been carried out to derive basic population counts for the 1971, 1981 and 1991 Censuses according to the 1991 community boundary set (Table 9.4). Figure 9.9 depicts the outcome of these operations by showing the location of the 392 centroids of the 1981 EDs, 150 (38.3%) of which were also 1971/1981 parishes, in respect of the 1991 communities, it indicates that only a minority (23 or 5.9%) of 1981 EDs had their centroid outside the 1991 community to which their population weighted tract centroid was assigned (solid dots). The shaded zones in Figure 9.9 are those parts of the 1971/1981 parishes lying outside the 1991 community. Although there were 102 such zones, many were relatively small in terms of area and hence apportioned population, it is clear that the larger ones (mean area 10.3 km^2) contained the centroids of 1981 EDs that were outside the 1991 community with which they were associated and this has been used as the criterion for reallocating these larger areas.

Figure 9.10 summarizes the extent of population change that occurred. The population change during the two periods (1971–81 and 1981–91) has been converted into a percentage index so that the mean in each case equals 100. The analysis indicates the relative importance of the change in the two decades. It is evident that

Table 9.4 Geocomputation procedures to produce comparable population counts for 1981 (1971) parishes and 1991 communities

1991 communities
• Determine population-weighted centroids for the 1971–81 census tracts (communities) – 1981 ED centroid used in the case of single ED tracts.
• Initially assign 1971 and 1981 parishes to 1991 communities by point-in-polygon searching of centroids and aggregate the 1971 and 1981 population counts.
• Assign 1971 and 1981 parishes to 1991 communities by polygon overlay of digital parish, MB and UD boundaries with the 1991 community boundaries and aggregate population counts.
• Reallocate parts of 1971/1981 parishes with their 1971 and 1981 population counts, if their own centroid falls in a different 1991 community to the one determined by polygon overlay and re-aggregate population counts as necessary.

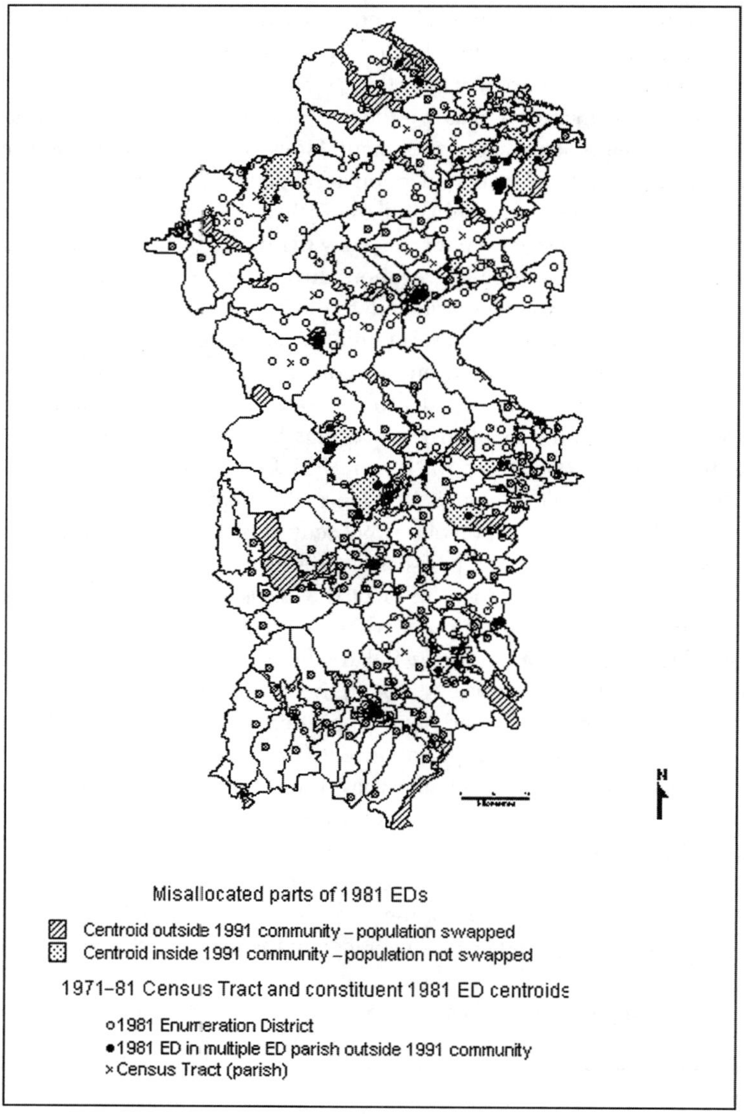

Figure 9.9 1971–81 parishes and 1981 enumeration districts in relation to 1991 communities in mid-Wales study area.

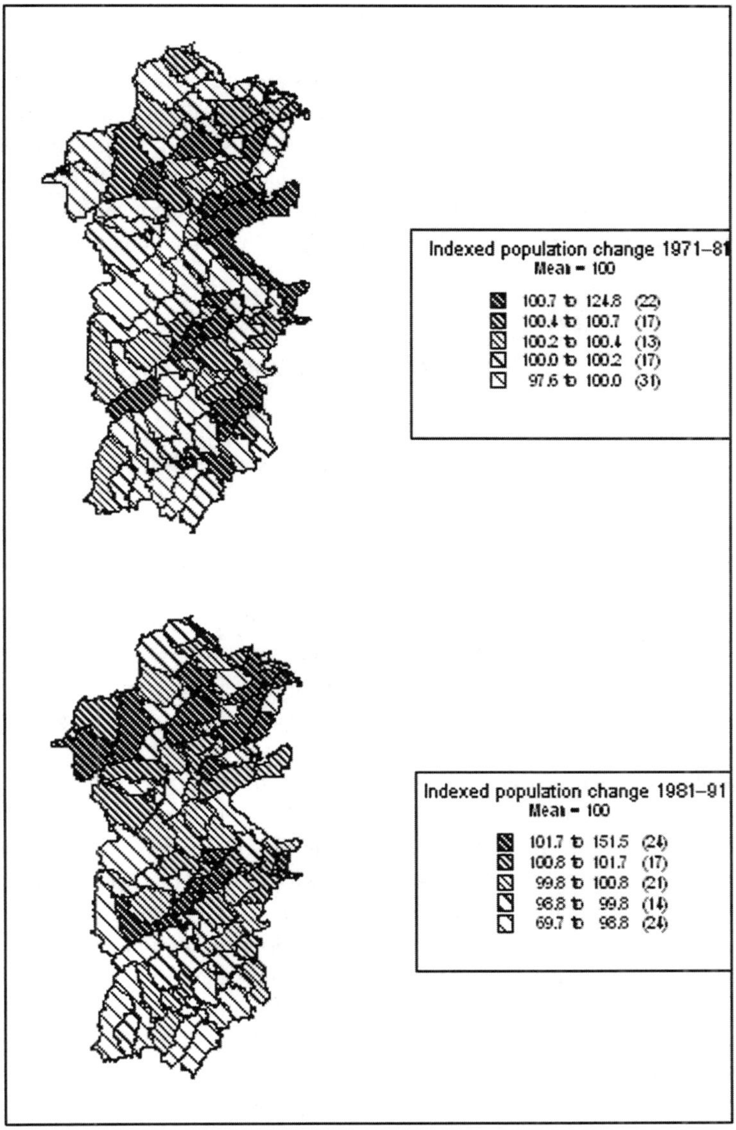

Figure 9.10 Summary of net population change and population totals in mid-Wales study area, 1971–91.

the communities experiencing the greatest growth in population between 1971 and 1981 tended to be on the eastern fringe of the study area close to the border with England. In contrast, larger population decreases mainly occurred in the central area and parts of the south. During the second decade, it would appear that population growth had become slightly more widespread, whereas the more extreme decreases continued in the southern part of the study area. Clearly this analysis has only hinted at the demographic changes that took place within this region and more detailed investigation would be required to discover the processes of change reflected by these aggregate figures.

9.6 PROBLEMS OF INTEGRATIVE GEOGRAPHICAL DATA ANALYSIS

It is readily apparent that there are many different sources of geographical data and that some are better suited to answering analysts' questions than others. However, an important unifying characteristic of the different datasets is the fact that geographical data refer to areas or places (geographical entities) on the Earth's surface. Geographical space constitutes an important organizing framework to which data entities can be referenced. Datasets that have been collected entirely independently of each other can be connected or integrated by virtue of the fact that they relate to the same part of the Earth. Unfortunately one of the main problems with trying to link together information for the same place is that the way in which data have been georeferenced and defined may vary.

The ability to integrate data for entities or from different sources that are spatially referenced to the Earth's surface has commonly been lauded as one of the key functions or virtues of GIS software. For example, in the late 1990s Intergraph's GeoMedia® Professional was described as 'tools for smart data capture, maintenance of maps and descriptive information, and *integrated* vector/raster maps' (emphasis added). Integration in this chapter refers to a process of manipulation or analysis that involves bringing together data in either one or a combination of two different ways. On the one hand, data of the same type (remotely sensed, cartographic, survey or administrative) that relates to different, but potentially related, thematic interests can be linked together to produce something that is 'greater than the sum of its parts'. On the other hand integration involves linking together data of two or more different types (i.e. from two or more of the four categories) about the same or different thematic topics with respect to a particular area or feature of geographical space. The essential difference between these two forms of integration is that the latter has the added complexity that fundamentally different types of geographical data are combined, whereas the former involves harmonizing thematic attribute information.

Figure 9.11 illustrates the first of these two forms of geographical data integration, where the essential aim is to find the 'right' answer to thematic questions about a particular place. In this case the questions being asked are how much land of a particular type exists in a given locality and where it is located. The example highlights the problems of attempting to integrate data sources that have different forms of georeferencing and spatial units. Figure 9.11a shows the results of a field survey of land utilization in a 4 km^2 study area where each plot of land on a base map with a scale of 1 : 25 000 has been classified into one of 12 land utilization categories (grassland, woodland, open space, heath/moor/rough land, transport, industry, settlement, water, arable, market gardening, unvegetated and derelict). The individual land parcels (e.g. fields or housing plots) are not depicted in the interests of clarity. In contrast, Figure 9.11b simulates the results of remotely sensed data collection where an image has been processed using procedures outlined in Chapter 4 into a raster grid of 50 m squares and these are categorized into the same 12 land utilization classes. When these two data sources relating to land use are combined or integrated, as illustrated at an enlarged scale for the northwestern quadrant in Figure 9.11c, it is evident that some points on the land surface are assigned to different categories by the datasets. Hence the question of what type of land use is to be found in these locations is answered ambiguously. For example, point A is grassland according to the field survey but heath, moor or rough land in the image data. Calculation of the extent of each land use for this section of the study area indicates there are 130 000 m^2 of grassland, 105 000 m^2 of heath, moor or rough land and 15 000 m^2 of arable. Whereas measurement from the field survey indicates corresponding figures of 120 463 m^2, 117 245 m^2 and 11 478 m^2 together with 814 m^2 of woodland. Table 9.5 compares similar area measurements for all land utilization categories across the study area from the two data sources, with those from the raster grid only accurate to the nearest 2500 m^2. In aggregate terms the differences may not be significant and will balance each other out in the sense that the total area remains 4 000 000 m^2

Table 9.5 Comparison of results of area calculations land use classes from field survey and raster image

	Field survey	Raster image	Difference	% Difference
Arable	170 812	147 500	23 312	13.6
Horticulture	5 970	7 500	−1 530	−25.6
Grassland	2 122 368	2 287 500	−165 132	−7.8
Woodland	117 508	142 500	−24 992	−21.3
Heath/Moor/Rough Land	834 900	840 000	−5 100	−0.6
Settlement	329 817	337 500	−7 683	−2.3
Open Space	39 670	0	39 670	100.0
Industry	232 741	157 500	75 241	32.3
Transport	120 843	65 000	55 843	46.2
Derelict	3 800	5 000	−1 200	−30.3
Unvegetated	15 543	0	15 543	100.0
Water/Marsh	6 028	10 000	−3 972	−66.4
Total Area	4 000 000 m^2	4 000 000 m^2	0	

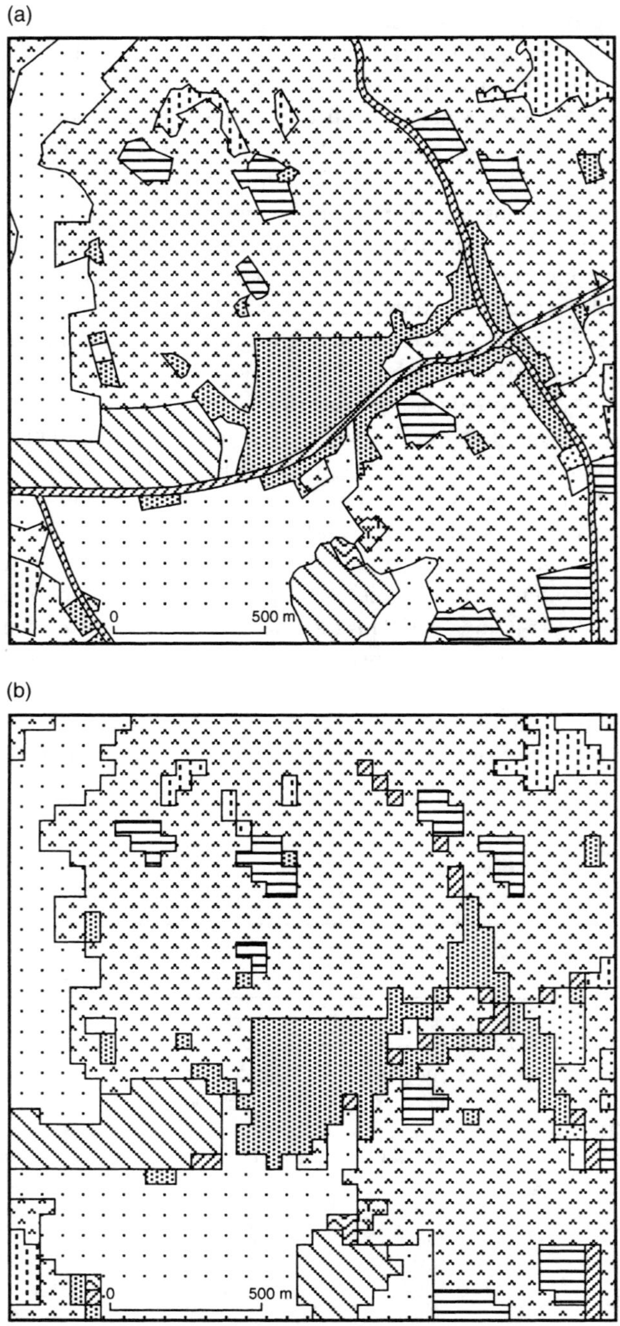

Figure 9.11 Integration of comparable thematic attributes from different types of data source.

(c)

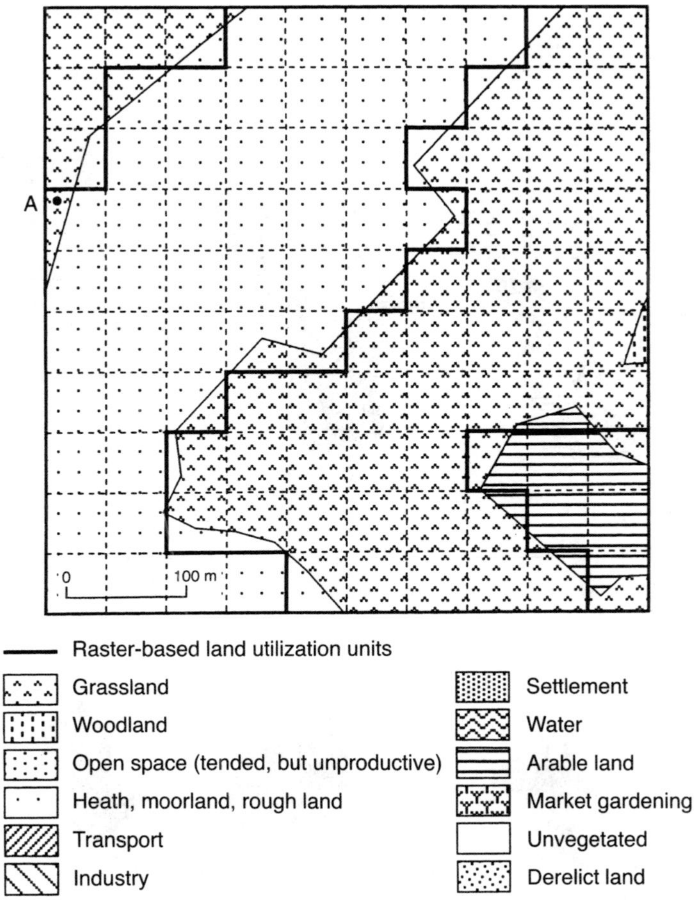

irrespective of the data source (Table 9.5). Nevertheless, it is clear that some land uses that occur in small units in relation to the grid size and/or are linear in shape tend to have been underestimated by the remotely sensed data producing the raster grid. A reduction in the size of the raster grid units would help to minimize these problems, for instance if 10 m squares were used the accuracy of the area measurements would increase to the nearest 100 m^2 and some of the originally misallocated locations would be categorized identically in the two data sources, but the origin of the inconsistencies would remain.

The second type of integration is illustrated in Figure 9.12, where the aim is to examine and possibly model the relationships between a range of thematic attributes for the same geographical features. The example shows three attributes calculated from Small Area Statistics from the 1991 UK Census of Population:

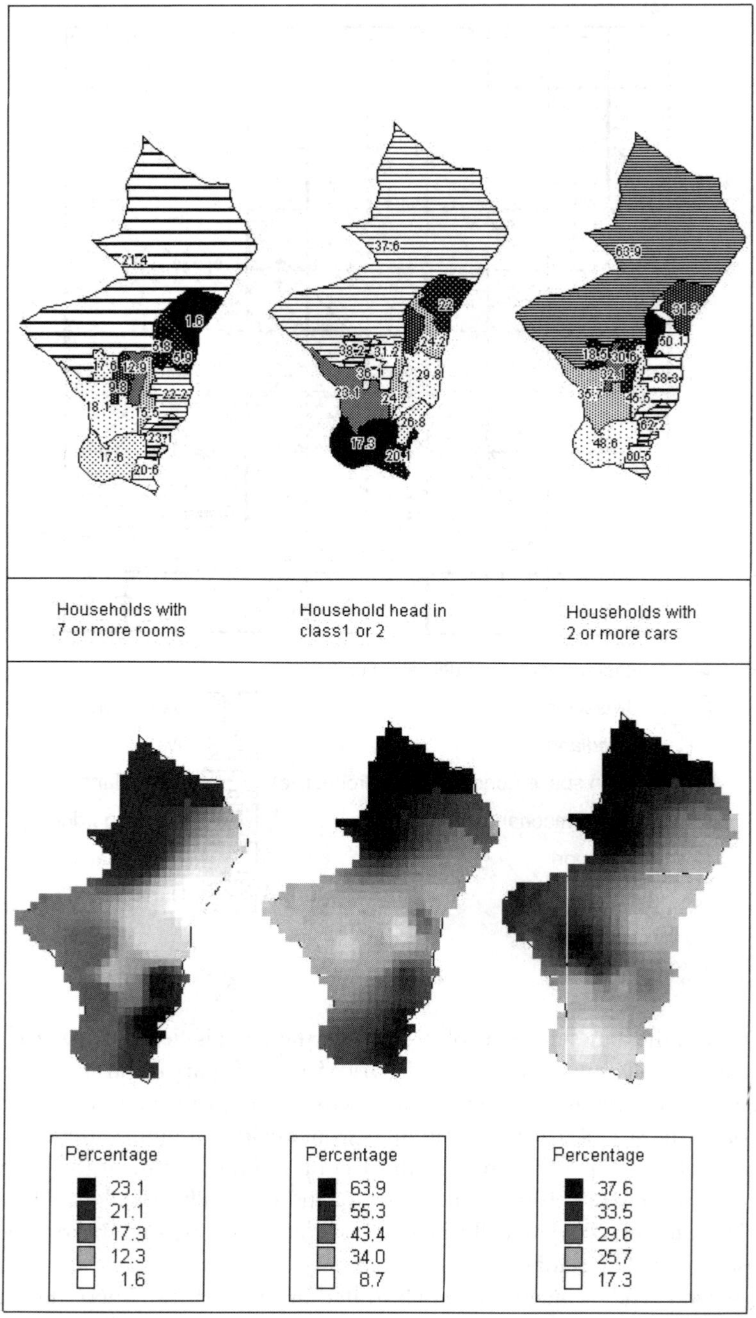

Figure 9.12 Integration of complementary attributes from the same data source. (Top) Individual attributes, (bottom) rasterized data.

- Percentage of households with seven or more rooms
- Percentage of households with access to two or more cars/vans
- Percentage of residents in households where head of household is in social class 1 or 2

The top part of Figure 9.12 shows the distribution of these attributes across the enumeration districts for the study area with the values shown as labels; the lower part shows the same information interpolated onto a 0.1 km grid and clipped to the boundary of the study area. A number of conclusions could be reached from an examination of patterns in each distribution. For example, residents in households whose head was in social classes 1 or 2 seem more likely to live in dwellings with seven or more rooms and to have access to two r more cars/vans. In contrast the smallest percentage for the social class and vehicle accessibility attributes occurs in the same enumeration district, which also has the second lowest value for households in larger accommodation. The grid-based distribution generally supports these ideas, but indicates very similar patterns for the dwelling size and social class attributes in the lower half of the study area, but a contrasting situation for households with access to two or more cars/vans. Further attributes could be added into the analysis and a matrix of correlation coefficients calculated between the different attributes. In this example, the only statistically significant Pearson's correlation coefficient when analysing the 13 enumeration district figures occurred between the percentages of households living in dwellings with seven or more rooms and of residents with household heads in social class 1 or 2 was 0.622.

Unfortunately these results may be misleading and the conclusions spurious. Irrespective of any violation of the assumptions of the correlation analysis, this example has disregarded two important problems facing the geographical data analyst – the ecological fallacy and modifiable areal units. Although the top part of Figure 9.12 clearly shows that 37.6% of households in the most northerly enumeration district had access to two or more cars/vans and 21.4% of households in this area had use of seven or more rooms, these two percentages did not necessarily overlap. It could have been the case that households with these characteristics were mutually exclusive, partially or completely overlapping. The available statistics do not tell us which is the case. In this example the three attributes originate from a single data source, the 1991 UK Population Census. However, the ecological fallacy problem can also arise when attributes from different sources for the same geographical units are integrated. The problem can only be satisfactorily overcome when it is possible to categorize the households or other phenomena in terms of two or more attributes in the raw data. The partitioning of geographical space into areal units, such as enumeration districts and grid squares, is essentially an arbitrary process. Boundaries of irregular area features may be drawn according to logical and practical rules, for example, that they follow recognizable physical features (e.g. canals, railway lines, roads) and that they contain a comparable number of the phenomena being quantified, but they nevertheless remain arbitrary in the sense that they could be defined differently and still satisfy the same criteria. The size of square units in the regular grid and the location of its origin are two elements that allow for a certain amount of arbitrariness in the units depicted in Figure 9.12.

There is a clear connection between data integration and the problems associated with examining spatial variation or temporal change that have been examined in the previous sections. Figure 9.13 shows how a series of small geographical units, which may themselves have either constant or changing boundaries, are capable of being switched between larger areas at different time periods (t, $t+1$, $t+2$, $t+3$, ... $t+n$). One example of where this occurs is in respect of parishes in England. The periodic redrawing of parliamentary constituency boundaries results in individual parishes being moved from one constituency and another (Youngs, 1979). Hence, if attempting to construct a time series of population information from the annual electoral registers, which are for the most part collected and organized on a parish within parliamentary constituency basis, it is important to be aware of discontinuities in the sequence. This characteristic represents a further permutation of the modifiable areal unit problem.

9.7 INTEGRATIVE ANALYSIS PROJECT

The aim of this project is to illustrate how data with inherently different characteristics, but relating to the same geographical area, can be combined. It examines the issues and problems associated with integrating traditional topographic mapping of boundaries with information obtained from scanned aerial photographic images. Using this geographical connection enables attributes and variables from the different data sources to be analysed together. The project has been kept relatively simple in order to demonstrate the principles of integrative analysis, which can become quite complex when a number of different datasets are used.

9.7.1 The Problem

In recent years the analysis of demographic and socio-economic statistics for small areas, such as unit postcodes or enumeration districts, often in conjunction with client or consumer information, has proved a valuable tool to assist with direct marketing, branch location and customer profiling. A number of commercial area classification products have been produced, including A Classification of Residential Neighbourhoods (ACORN) and CCN's MOSAIC system, and undoubtedly these will be updated as the latest round of Population Census statistics become available. However, such classification systems can be criticized for failing to take into account the environmental characteristics of the places where people live, or for assuming the built environment of their housing is of principal importance. Nevertheless, the descriptions of these places are constructed using such terms as 'Middle Class suburbs' or 'Agricultural Villages'. The aim of this project is to outline the feasibility of incorporating information about the environmental characteristics of the areas in which the various household classes live. These ideas are developed with respect to a

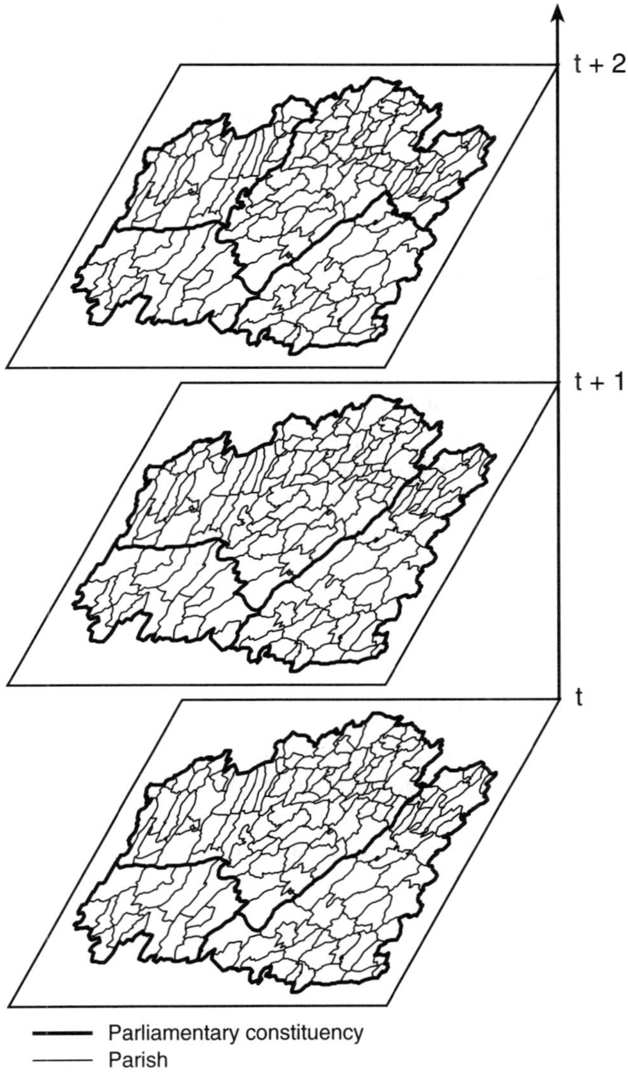

Figure 9.13 Modifiable areal unit problems associated with the integration of geographical data in space and time.

relatively small case study area, but one in which there are clear contrasts in the spaces surrounding dwellings in different parts of the settlement. The old core of the settlement dating back several centuries has narrow irregular streets with a mixture of small and large properties, Victorian and Edwardian housing occurs around this core and in small outliers, formerly isolated but now part of the semi-continuous built-up area. Inter-war and mid-20th-century estate developments tend to have more space, while much of the newest estate housing comprises dwellings at relatively high densities. The aim of the project is to determine whether these types of housing are not only differentiated in terms of the socio-economic and demographic characteristics of the people living in them, but also in respect of their environmental quality, as measured in this case in terms of the quantity of open space in the vicinity of the properties.

9.7.2 Data Sources

The two main types of basic data used in this project are counts from the UK 1991 Census of Population and aerial photographic imagery for the case study area taken in the summer of that year. A total of 128 'raw' census counts have been combined to produce the same set of 'indicator' variables that were chosen by ONS when undertaking a classificatory analysis of local authority districts (ONS, 1997) as detailed in Table 9.6. These have been extracted from the Casweb system (see Chapter 8) for all enumeration districts in one local authority together with the digital boundary file. The detailed analysis focuses on a single ward containing 13 EDs within this local authority for which aerial photographs were obtained with a nominal scale of 1 : 3000. For copyright reasons images of the original aerial photographs are not reproduced here, although an interpretation of their information is included.

9.7.3 Analysis and Results

The analysis undertaken for this project divides into three parts: multivariate statistical classification of the EDs to produce clusters or groups and the transfer of the results into MapInfo; the interpretation of the scanned aerial photographs and creation of polygons representing different environmental areas; and integrating and mapping the results of these analyses. A similar classificatory analysis has been carried out with respect to the EDs in the local authority to that originally undertaken for all districts in England and Wales by ONS (ONS, 1997). Given the particularity of its location, the same set of clusters is unlikely to arise and hence no attempt is made to compare the two sets of results. Cluster analysis comprises a suite of techniques that are generally used to collapse a relatively large number of entities into a smaller number of groups or clusters. It is based on the notion that entities can exist in multi-dimensional space that takes its metric from the collection of variables with respect to which the entities have been quantified. One of the most commonly used forms of clustering in geographical applications adopts the agglomerative approach by starting from the position that each entity is a cluster with one member and the analysis iterates $n-1$ times, where n = number of original entities. The number of clusters reduces by one at each iteration as two entities join together, or as previously formed clusters amalgamate. The 'decision' on which clusters join at each stage is made by reference to a proximity or distance measure, which is recalculated at the start of each iteration to take into account the location of the newly formed cluster. Cluster analysis is not a precise statistical procedure in the sense that more than one

Table 9.6 Variables derived from 1991 Census of Population Small Area Statistics counts used in cluster analysis

Variable Name	Description
ZONEID	
RES04	% residents 0–4
RES514	% residents 5–14
RES2544	% residents 25–44
RES4564	% residents 45–64
RES65	% residents over 65
RESBLACK	% residents identifying as black
RESASIAN	% residents identifying as Asian
HSSIZE	Average household residents per household
HS4KIDS	Households with 4+ children as % of households with children
SINGWKAG	% households which are single working age person households
HSOWNOC	% residents in owner occupied accommodation
HSPUBRT	% residents in public sector rented accommodation
HSPRIRT	% residents in private rented accommodation
NOCENTH	% residents with no central heating
DWTERR	% dwellings that are terraced
DWFLATS	% dwellings that are purpose-built flats
DW7RMS	% households in dwellings with 7+ rooms
RMSPP	Average rooms per person
MIGRNTS	% residents >1 living that are 1 year migrants
RESQUALS	% residents aged 18+ with higher qualification
HHCLS12	% household residents where head in class 1 or 2
HHCLS3	% household residents where head in class 3 (non-manual)
HHCLS45	% household residents where head in class 4 or 5
HSWKAGAD	% households with 2 working adults and no children
HS2CARS	% households with 2 or more cars
DEPLNCRE	% dependents in lone carer households
KIDS1AD	% children in single adult households
RESNOCAR	% household residents in households without a car
COMPUBTR	% persons in employment commuting by public transport
LGTMILL	Directly standardized rate for limiting long-term illness
UNEMPLT	Unemployment rate (%) for men and women of working age
WKGWOM	% working age women employed
INACSTUD	% residents who are economically inactive students
AGRICWK	% persons in employment working in agriculture
PRPRODWK	% persons in employment working in primary production
MANUWK	% persons in employment working in manufacturing
SERVWK	% persons in employment working in services

outcome is possible, since the analyst is presented with a number of options, for instance, about which distance criteria to use and how many clusters provide a satisfactory final solution.

In the present analysis the variables have been standardized to fall within the range −1 to +1 in order to even out variations in the value ranges. Squared Euclidean distance has been used as the similarity criterion in Ward's clustering procedure. A plot of the criterion values at which the clusters formed by the last 30 iterations

(Figure 9.14) shows that the gradient of the curve starts to increase more rapidly at about seven clusters and this number provides a realistic end point to the analysis. The next stage involves describing and labelling the clusters in some meaningful way by reference to their summary values on the original variables. Table 9.7 provides the means of the 37 variables for each cluster and for all EDs with values markedly above and below the overall mean highlighted in bold. Evidently there is some noticeable deviation away from the mean for most variables, which has provided the basis for assigning labels to the clusters (Table 9.8). Each ED has therefore been assigned to one of the seven clusters and the records from the statistical analysis file, once saved in Excel format, have been joined with the ED digital boundary table stored in MapInfo using the standard area code as the indexing attribute. Figure 9.15 maps the outcome of connecting the 'cluster membership' variable to the appropriate polygons and highlights the study area ward where three clusters dominate.

The vertical aerial photographs for the single ward within the local authority were scanned at 600 dpi (dots per inch) and then combined to produce a mosaic of the study area. For the present purpose this raster image has not been processed using techniques outlined in Chapter 4, it has been read into MapInfo and registered to British National Grid co-ordinates. Four types of environmental area were defined by visual interpretation of the combined image – houses, gardens and other buildings, woodland, open space and agriculture – and polygons for these were digitized directly

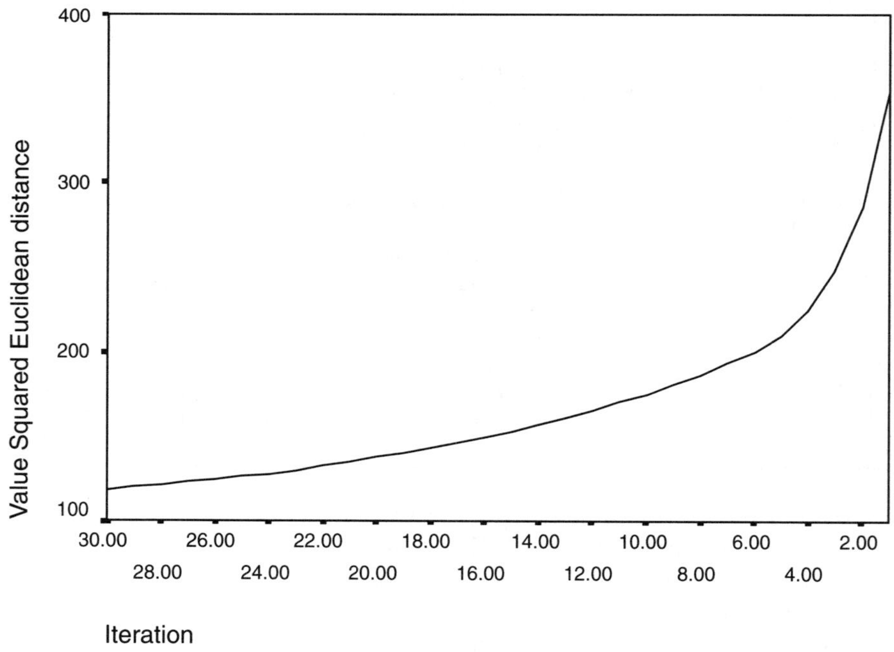

Figure 9.14 Scree plot from cluster analysis.

onto the displayed image. The analysis involved overlaying the cluster group and environmental area maps in order to calculate the extent of each type within the EDs. This process has enabled additional variables to be calculated for each enumeration district relating the demographic information to the extent of each environmental

Table 9.7 Summary statistics (means) for seven clusters and all enumeration districts

	Cluster							
Variable Name	1	2	3	4	5	6	7	Total
RES04	6.3	5.5	8.6	6.2	**0.3**	4.0	**16.3**	6.3
RES514	7.9	11.4	14.1	13.6	**0.3**	11.5	**18.0**	11.7
RES2544	35.0	28.1	28.5	30.1	25.2	**24.3**	**42.7**	29.8
RES4564	*16.5*	23.9	19.4	23.3	5.8	**28.4**	**2.7**	20.9
RES65	16.3	17.1	14.8	13.6	14.9	**17.6**	**1.0**	15.2
RESBLACK	0.9	0.4	0.8	0.3	**2.8**	**0.1**	0.7	0.6
RESASIAN	**0.9**	0.5	0.5	0.4	**0.9**	**0.3**	**0.3**	0.5
HSSIZE	2.1	2.5	2.6	2.7	**0.1**	2.6	**3.1**	2.4
HS4KIDS	2.4	2.2	**5.0**	3.1	**0.0**	4.3	2.2	3.0
SINGWKAG	**21.7**	9.2	9.7	6.9	**0.3**	8.5	*3.9*	10.2
HSOWNOC	75.4	86.9	47.4	**88.2**	**1.2**	85.2	8.6	73.2
HSPUBRT	6.5	7.4	**45.6**	5.3	**0.0**	5.9	3.5	12.5
HSPRIRT	**12.4**	4.2	2.9	3.7	**0.3**	5.8	2.9	5.1
NOCENTH	**18.6**	6.3	10.6	5.2	**0.6**	7.5	4.7	8.6
DWTERR	42.6	10.8	39.1	6.0	**0.4**	5.3	**57.0**	20.0
DWFLATS	19.1	12.0	**27.6**	**3.4**	6.8	5.5	17.9	12.9
DW7RMS	7.9	17.9	6.6	29.2	**0.8**	**39.5**	15.4	18.6
RMSPP	2.2	2.2	1.8	2.2	**1.6**	**2.5**	**1.6**	2.1
MIGRNTS	19.8	9.3	10.8	**7.7**	38.0	8.0	**39.6**	12.7
RESQUALS	17.5	14.6	**4.9**	15.1	9.3	**25.7**	6.5	14.2
HHCLS12	39.5	42.2	18.6	48.5	**0.0**	**75.3**	5.3	39.8
HHCLS3	13.0	**17.6**	10.6	9.8	**0.0**	6.1	4.2	11.6
HHCLS45	22.3	11.8	**33.0**	12.4	**0.0**	6.1	4.2	16.0
HSWKAGAd	0.1	0.2	**0.9**	0.3	**0.0**	0.1	0.1	0.3
HS2CARS	15.1	28.7	16.9	42.7	*0.1*	**44.6**	9.2	28.0
DEPLNCRE	21.2	15.5	**24.4**	13.8	**0.5**	15.3	7.0	16.6
KIDS1AD	17.2	8.0	**21.0**	6.9	**0.0**	7.4	5.1	10.9
RESNOCAR	27.9	13.8	**31.8**	7.5	**6.2**	9.0	21.6	17.0
COMPUBTR	13.4	13.7	**19.4**	11.8	**0.0**	12.7	5.4	13.3
LGTMILL	10.4	8.6	**12.3**	8.1	4.6	8.0	**2.5**	9.0
UMEMPLT	9.3	5.7	**12.9**	**4.9**	6.0	6.6	7.1	7.4
WKGWOM	**65.8**	61.8	54.4	59.1	**37.0**	53.9	49.0	58.5
INACSTUD	3.7	4.1	3.0	4.3	**20.1**	5.7	**2.0**	4.6
AGRICWK	1.9	1.2	1.6	3.1	**0.0**	**8.9**	**0.0**	2.5
PRPRODWK	1.6	**2.8**	2.1	1.2	**0.0**	**0.0**	0.4	1.7
MANUWK	12.2	11.7	15.0	**18.2**	**0.0**	7.3	5.6	13.0
SERVWK	47.6	44.7	**34.8**	41.0	54.6	53.5	**78.8**	44.7
N of EDs	49	77	50	78	11	27	9	301

Note: Highest and lowest means for each variable shown in bold.

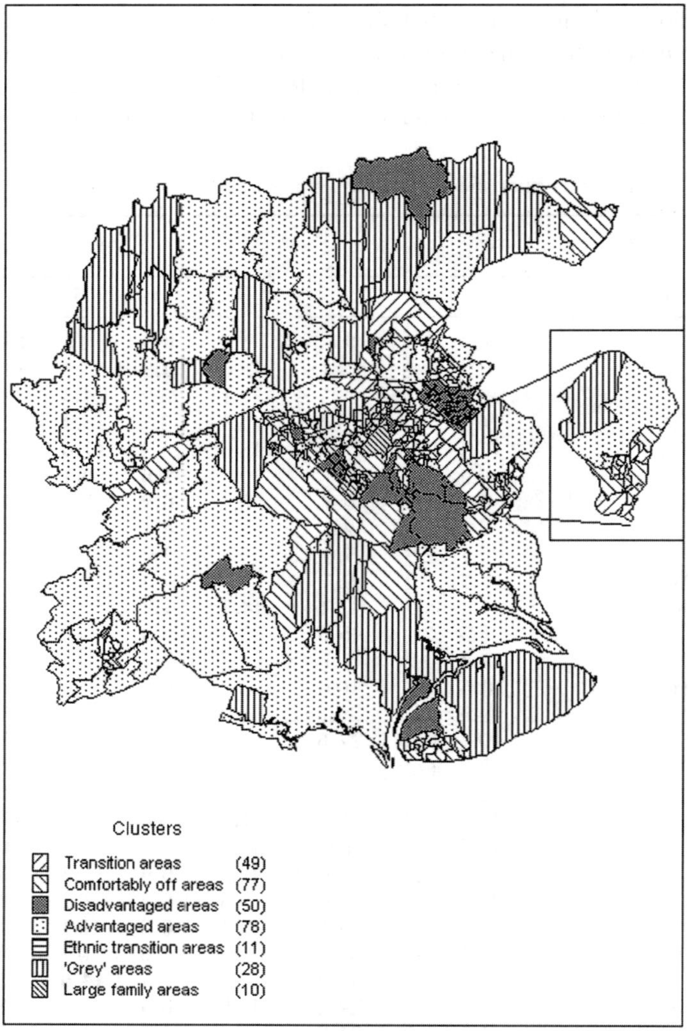

Figure 9.15 Distribution of clusters across case study local authority.

type. The results of this analysis not only provide an improved measurement of such standard indicators as population or housing density by referring to the built-up area, but also enable variables such as distance to woodland, open space or farmed countryside, and the amounts of these per person or household to be assessed. Table 9.9 summarizes the results of this analysis on EDs by cluster type. The limited number of EDs involved in this analysis inhibits statistical analysis of the results, nevertheless it is apparent that there was a marked difference in the number of persons per hectare of built land across clusters types 1, 2 and 4, although the number of households per

Table 9.8 Labelling of clusters

	N of EDs	Cluster label
Cluster 1	49	Transition areas: Young adults and single persons in privately rented terraces and flats without children or cars
Cluster 2	77	Well off areas: Middle class households in owner occupied housing
Cluster 3	50	Deprived areas: Working class households in local authority flats and terraced housing with unemployment and a lack of private transport
Cluster 4	78	Better off areas: Upper-middle class households in large, owner occupied houses with car and working in manufacturing sector
Cluster 5	11	Ethnic transition areas: Persons of non-white ethnic origin with migrants and economically inactive students
Cluster 6	27	'Grey' areas: Older adults and pensioners in upper social classes in large owner occupied housing
Cluster 7	9	Large family areas: Large sized households with young adults and children in terraces and flats

Table 9.9 Comparison of clusters with environmental type variables (means)

Cluster	No. of EDs	Built land	Woodland	Open space	Agriculture	Persons per ha built land	Households per ha built land
1	4	10.0	1.0	6.9	0.0	43.8	20.3
2	7	10.3	3.3	3.2	2.6	55.2	21.7
4	2	7.7	7.8	8.0	144.2	63.5	21.4

Note: The single EDs in clusters 5 and 6 have been excluded.

hectare was similar in each case. The relatively 'well-off' EDs in cluster 4 enjoyed on average more woodland and open space than cluster 2 and they may be considered semi-rural on account of the amount of agricultural land they contain. This project may reasonably be regarded as a pilot study which would enable more refined environmental variables to be derived, for example, distinguishing between the area of dwellings and gardens within the built land category. These and other environmental indicators could be combined with those counts originating from the census in order to produce a new 'enviro-demographic' classification.

Chapter 10

Future Developments in Geographical Data

10.1 INTRODUCTION

In many respects there is no ideal time to start writing the concluding chapter of a book about the characteristics and accessibility of sources of geographical data. As little as 5 years ago, when the idea for this book originated, the situation in both respects was very different. The presence of georeferencing or spatial coding in data, in other words 'geography', has become increasingly evident in recent years. The role of space as a framework for organizing data about phenomena has become more important as the advantages of visualizing information in a cartographic fashion have been realized by a wider range of organizations. It is surely no coincidence that these changes have occurred at a time when there is a common perception that the 'world has become a smaller place'. It has been important to know how much of what is where since societies formed in relatively confined and local spaces during the early stages of civilization. However, it seems reasonable to argue that such knowledge has become even more important as we realize our capacity to witness events on the other side of the globe as they are occurring.

The BBC's coverage of the start of the new millennium aptly illustrates this ability. At 10.00 h GMT on 31 December 1999 the British television audience was able to watch a live broadcast of the first sunrise of 2000 on Millennium Island (formerly Caroline Island) in the Republic of Kiribati. This island is located at 10° 00′ S and 150° 14′ W lying about 3600 km south of Honolulu, 900 km north of Tahiti and 6100 km southwest of San Francisco (Figure 10.1). The BBC coverage of this event lasted over 24 hours and tracked across the Earth as successive countries awoke to the dawn of the new millennium. By the time that the bells of Big Ben in London were chiming midnight on 31 December 1999, the Pacific islanders were already more

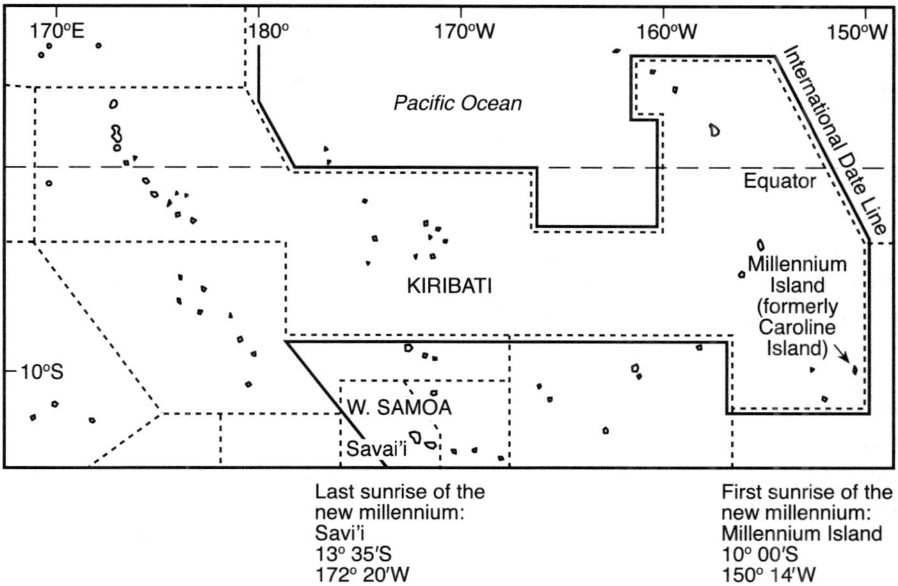

Figure 10.1 Pacific Islands experiencing first and last sunrise of 1 January 2001 in relation to International Date Line.

than 12 hours into 2000. The last place on Earth to enter the new millennium was the largest Samoan island of Savai'i located at 13° 35′ S and 172° 20′ W, which places it further west than the slightly more northerly Millennium Island that had witnessed the first sunrise of 2000 some 24 hours earlier. This example reminds us not only of the importance of the passage of time, but also the intimate interconnection between space and time in our daily lives. Many people saw the new millennium as an occasion for celebration and for making a new start. But how long was it before the routine and normality of everyday life returned and it seemed to make little difference that it was 2000 rather than 1999?

The high profile of this example highlights our perception of the connection between space and time, but for many people the ability to travel around the Earth on business or to visit friends or relatives is now an accepted part of life. The corollary of this broadening of spatial or geographical awareness is that, as knowledge of and contact with more distant parts of the Earth have become commonplace for a greater number of people, so awareness of our own locality, of the things in our immediate vicinity, have taken on a greater significance. In respect of geographical data this has led to a desire for more detailed or spatially specific information, if not in place of, at least as a foundation for, the broad brush statement. It is inevitably difficult, perhaps even futile, to attempt to unravel the complex chains of causality connecting technological progress in ICT, the emergence and continued development of GIS and the greater use of georeferencing and spatial coding for organizing data about terrestrial phenomena. It is clear that GIS and computer mapping have now become a commonplace means of analysing and displaying spatial data. Whether these operations will become as normal as using database, spreadsheet and word-processing software is still open to debate. However, it is clear that one factor contributing to the wider use of these spatial information handling technologies is the increased availability and packaging of geographical data.

All this may seem somewhat removed from the consideration of the four categories of geographical data examined in this book, of the various examples used to illustrate their characteristics and the examination of geographical data coverage for a selection of countries provided in Chapter 8. Five years ago the various types of data, remotely sensed, cartographic, census or survey derived and administrative/statutory, were all produced and made available as secondary sources to analysts. However, during the intervening period, a number of developments have occurred that have changed the ways in which people can access data. Principal amongst these are the progressive spread of the internet and a greater willingness in organizations, particularly in the public sector, to make their data available to other people. Cynics might argue that we are still only being allowed access to a limited range of information, which may have had data values rounded to base 3, 5, 10, 50 or to an even higher level of generalization, even if these data are for smaller spatial units than previously. The speed of change in the recent past advises caution when trying to reflect on possible future directions in geographical data sources, nevertheless this concluding chapter considers some of the current developments and thereby highlights possible futures.

10.2 WEB-BASED MAPPING

The principle underlying this book has been that users of GIS and computer mapping software will want to combine spatial and geographical information in order to undertake some form of spatial analysis or visualization. It has emphasized the distinction between geographical data relating to the spatial or geometrical properties of terrestrial phenomena and those representing attributes or variables that measure or categorize these features. In other words it was predicated on the notion that those wishing to produce maps for illustrative purposes would need to experience the pain of searching out the requisite data sources, manipulating these in such a fashion that they were compatible and then designing the map or other graphic device and, finally, outputting this on hard or soft copy media for display. It is undoubtedly the case that for the foreseeable future this general sequence of events will be followed by those working at the research end of the GIS/computer mapping user spectrum. However, for those 'simply' wishing to incorporate some form of geographical illustration in material that is not specifically concerned with the spatial characteristics of the phenomena, but only in the thematic distribution attached to them, may avail themselves of facilities on the internet for Web-based mapping. Some other areas of GIS functionality may be present in Internet GIS (IGIS), such as spatial searching, distance determination or network navigation, although open access to the range of facilities in 'full-scale' systems is still some way off.

Figure 10.2 illustrates a simple model of IGIS architecture based on the client/server principle. This model involves two programs communicating with each other across a computer network, either a Local or Wide Area Network (intranet or internet), using an established protocol language, usually Hyper Text Transfer Protocol (HTTP). The two sides of the model are linked by 'middleware' programs that break down the information distributed from the server side into small packets and checks that these are received correctly by the client machines (PCs). The client side of the model uses a web browser to select the desired host server and to present the packets of information mediated through the 'middleware'. An IGIS usually incorporates a standard Web server and a GIS application server where the geographical data are stored. An alternative approach involves the GIS functionality residing on the client side machines using programs such as Java or ActiveX in conjunction with the browser software to produce specific spatial analysis operations. Harder (1998) suggested that geographic information on the internet can be used to produce relatively simple maps as well to undertake more complex spatial analyses.

Many people already use free IGIS services to obtain maps for locational and navigational purposes. There are a number of competing service providers, most of which provide free access to street maps of the UK and a variable number of other countries. The usefulness of these maps inevitably depends on the underlying accuracy of the geographical databases underpinning the service. For example, the level of precision used to georeference UK postcodes (100 m) means that these will not locate the exact address that has been entered. Table 10.1 provides a summary of the services and GIS operations contained in a selection of these systems in the UK. Most

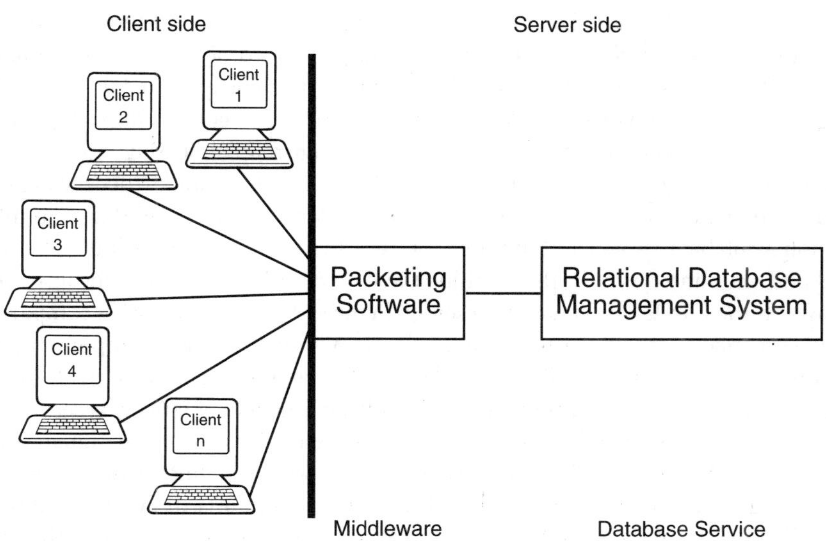

Figure 10.2 Client–server architecture for internet Geographical Information System.

Table 10.1 Comparison of GIS interrogation functionality at selected commercial internet mapping sites

Provider	Services	GIS functions
www.multimap.com	UK street maps	Locate places via UK postcodes and/or addresses or map pointer.
	Travel directions	Route planning using postcodes and/or addresses.
	Local information	Reciprocal links to categories of information via place name, postcode or London street.
www.mapquest.co.uk	UK street maps Travel directions	Locate places via UK postcodes, addresses and/or latitude–longitude. Route planning using postcodes and/or addresses.
www.ordnancesurvey.co.uk www.streetmap.co.uk	UK street maps	Locate places via UK postcodes, addresses, OS grid reference, telephone area code and more.
www.getmapping.co.uk	Aerial photographs	Locate places via UK postcodes and/or addresses.
www.upyourstreet.com	Location map and local information	Locates places by postcode and find out property prices, nearest facilities, events, ACORN® profile, council performance and much more.

offer a search facility based on either entering a postcode and/or address or in some cases pointing at a UK (Great Britain) map display. Two of the providers listed in Table 10.1 also allow route planning by the shortest or cheapest means, identifying origin and destination by reference to addresses and/or postcodes. One case provides access to aerial photographs of Britain with a coverage map indicating availability. In all cases the map graphics are displayed using the raster model and hence may appear somewhat jagged. These providers are clearly making geographical data and information available to a wider community of users than was possible as little as 5 years ago, and undoubtedly many people using these services are unaware of the term GIS. However, the GIS functionality represented in the systems is restricted in comparison to what regular users of GIS are familiar with. However, perhaps the important point to consider is that these facilities now allow people to ask and get answers to simple geographical questions in a way that could only be imagined relatively recently.

The next generation of Web-based mapping systems allow users to design maps for themselves. One such system, know as Descartes, already augments the facilities on offer at MIMAS (see Chapter 8). This is a tool that supports the visual analysis of spatially referenced data and has been developed by the German National Research Centre for Information Technology with the company Dialogis. Descartes has been made available from MIMAS to registered users with ward level statistics from the 1991 UK Population Census and to unregistered users with a demonstration dataset for Leicestershire. The essential features of the system are that users can select column(s) of data (e.g. overcrowding and limiting long-term illness) and then visualize this information interactively over the internet. The visualization may be preceded by user-controlled classification of the raw data to generate one or more maps, to look for relationships between the values in two or more columns or to produce statistical graphics (e.g. column and pie charts). Users can interact with the display to pan and zoom the map, to select features and to change the design. Descartes acts as an interactive data exploration tool enabling users to experiment with various types of visualization.

10.3 INTERACTIVE DATA ACCESS

There is often a tacit presumption that the analysis or interrogation of geographical databases will inevitably involve the display of the resultant information as maps. However, this is clearly not a prerequisite and a number of systems have been developed that provide access to and display of geographical statistics as tables, graphs or simply text. There is a distinction between those that enable users to download the desired raw (unprocessed) data for further analysis and others that allow users to consult the databases in order to obtain answers to questions. Some of the sections in Chapter 8 briefly reviewed systems set up in different countries to provide access to metadata about geographical datasets and to raw data that includes georeferencing or spatial coding. Some of the latter are generally available, such as the facility for downloading small-scale digital mapping files from the USGS and free maps that can be obtained from the OS, whereas others involve user authorization, since they are designed for a particular client base (e.g. the Casweb service at MIMAS) or charge users for access.

In response to the demand for access to geographically specific information from surveys and censuses collected by national governments, the Statistics Departments in a number of countries have recently established services that provide access to the results of their Population and/or Agricultural Censuses. The Population Census (*Recensement de la Population*) section of the French INSEE website enables anyone to search out some basic population count figures for any commune in France simply by typing its name. A similar system exists with respect to the 2000 Census of Agriculture for France, although, at the time of writing, the data are only available online for 20 *départements*. It is possible to discover the following statistics for *communes* within these areas:

- Total number of farms
- Number of professional (commercial) farms
- Number of farmers and partners
- Number of farm family members engaged in agriculture

- Total number of persons engaged in agriculture (full-time equivalents)
- Area of land on farms
- Agricultural utilizable area
- Permanent grassland
- Total number of cows
- Total number of farms in previous census (1988)

As an example, Table 10.2 illustrates the outcome of such an internet-based interrogation for the *commune* of *Ardes* in *Pas de Calais*. The information provided represents only a predetermined selection of counts from the full agricultural census database. A somewhat different example, but still relating to France, is provided by the IGN Website which allows users to find out the georeference of *communes* throughout France in terms of latitude and longitude. In New Zealand a similar system to those developed for the French censuses has been established with respect to the provisional counts from its Census of Population. The latest 'census day' in New Zealand was Tuesday, 6 March 2001 and by late May 2001 it was possible to obtain preliminary dwelling and population figures at unit area level throughout the country and to download these as raw data files in Comma Separated Variable (CSV) format for use in analysis software. The unit areas are non-administrative aggregations of meshblocks smaller than territorial authorities and are single geographic entities with a unique name.

It will take somewhat longer for results from the UK's 2001 Census of Population to be incorporated in the fledgling Neighbourhood Statistics Service that was launched on 26 February 2001. Although the extent of census output that will be included in this collection of statistics is still to be decided, there is already a commitment to providing at least a range of indicators for the new Output Areas and wards. As these examples illustrate, individuals and organizations interested in analysing geographical data or at least in being able to answer relatively simple questions about the

Table 10.2 Sample output of enquiry commune level figures from the 2000 French Agricultural Census

Département: **PAS-DE-CALAIS**	Commune: **ARDRES**
• Nombre d'exploitations	12
dont nombre d'exploitations professionnelles	9
• Nombre de chefs d'exploitation et de coexploitants	15
• Nombre d'actifs familiaux sur les exploitations	35 personnes
• Nombre total d'actifs sur les exploitations	19 UTA (équivalent temps plein)
• Superficie agricole utilisée des exploitations	482 ha
• Terres labourables	431 ha
• Superficie toujours en herbe	51 ha
• Nombre total de vaches	110
• Rappel: Nombre d'exploitations en 1988	26

Source: Recensement Agricole, France 2000.

characteristics of small areas now have a greater opportunity to do so than in the past. Assuming that these developments will be used and at least partially satisfy the latent demand that has been identified, one can only envisage that such systems will mature so that user interfaces become more intuitive and wider ranges of information are included.

10.4 NEW SOURCES OF GEOGRAPHICAL DATA

The reviews of the four categories of geographical data sources in Chapters 4 to 7 have revealed that, although some datasets are idiosyncratic or 'one-offs', in many cases they form part of a series and once initiated they are soon repeated with data collection occurring with relatively more or less frequency. The time interval between collection events may be as long as 10 years, as in the case of most population censuses, as little as every day, for example with the registration of births and deaths, or even shorter, as with the hourly monitoring of environmental and meteorological conditions by satellites or ground-based systems. So it seems clear that there will be an abundance of geographical data relating to the features of the Earth's surface going forward into the indefinite future.

In essence these supposedly new datasets will simply be updates of what has been collected previously, in other words they will be new versions of old datasets. They will tell us about the new spatial configuration of geographical features, for example the new location of a coastline, or about the new characteristics of schools, for example the number of pupils with special educational needs. These new datasets provide the opportunity to investigate change and to examine the most up-to-date information. Genuinely new datasets, which may be defined as those collected using an innovative type of measuring device or produced as a result of a novel merger of information from existing sources, are less common. A recent example of a new measuring instrument is the Light Detection and Ranging (LIDAR) airborne remote sensing device that uses laser light in a similar way to radar uses radio waves and sonar uses sound. Hence rather than capturing EMR emanating passively from the Earth's surface, as outlined in Chapter 4, LIDAR is based on the principle that by emitting a pulse of laser light the time and quantity returning to the instrument can be recorded and, knowing the speed of light, can be converted into height. The Vegetation Canopy Lidar (VCL), part of NASA's Earth System Science Pathfinder programme, has the aim of creating a three-dimensional representation of the Earth's structure by vertical and horizontal measurement of canopy and land surface topography.

Remote sensing also provides the second example of how a genuinely new type of geographical dataset might be collected. Some 40 million mobile telephones are registered within the UK alone and the majority of these are likely to be switched on and about the person of the user during waking hours in anticipation of receiving a call or text message. This means that the location of such a telephone, hence the user, can be determined from its signal and so the person's movements can be tracked.

This adaptation has the potential to revolutionize how some types of information are collected. Instead of asking people in a survey how they travel to work, it would be possible to track people's movement and changes of transport mode via their telephones and to link this to existing digital map databases. Hence, if the telephone habitually moved along the co-ordinates of a known railway feature between 07.30 and 08.20 h between Swindon and London, then it could reasonably be inferred that the owner was travelling to work by train. As with most datasets, this system might have some inaccuracies, so that the purpose of any individual journey may not be determined. For example, if the person happened to be on annual leave one day and decided to make the same journey for leisure or social reasons, then this would not be recorded, although this could perhaps be inferred if subsequent movements indicated visits to a cinema, museum or theatre rather than to the usual place of work grid reference. Similarly there may be interruptions to the continuous flow of data when people turned their telephone off, when the battery had discharged or when the signal was impeded. Apart from these problems, this system would not necessarily provide a complete census of who was travelling along a particular railway track since there are still some people who have eschewed obtaining a mobile telephone, thus far.

It is perhaps questionable whether a dataset produced as a result of merging two existing data sources in a novel way genuinely comprises 'new' geographical data, as opposed to simply a new GIS application. The importance of postal geography in the UK and elsewhere as a means of targeting and identifying where people live has been examined in a number of places in previous chapters. One recent development by the Royal Mail has seen the combination of address postcodes with information held in the land registry enabling the names and addresses of first-time buyers of residential properties to be identified. This development within the field of geodemographics is likely to prove useful to firms selling products that new householders may need, or be persuaded that they need, to purchase. The example also illustrates how there is an intimate and mutually dependent connection between the data sources used in GIS and the spatial analysis and applications that these systems can perform. This dependency suggests that innovative spatial analytic procedures will continue to be developed, even if some of the more mundane functionality of GIS becomes routine.

Bibliography

Allen, J. A. (1984) The role and future of remote sensing, in Remote Sensing Society, *Satellite Remote Sensing – Review and Preview*, Proceedings of the Tenth Anniversary International Conference of the Remote Sensing Society, University of Nottingham.

Alonson, W. and Star, P. (eds) (1987) *The Politics of Numbers*, Russell Sage Foundation, New York.

Antenucci, J. C. (1995) Rasters and attributes. *Geographic Information Systems*, **5(2)**, 44–45.

Aranha, J. T. M. (1998) *An Integrated Geographical Information System for the Vale do Alto Tamega (GISVAT)*, PhD Thesis, Kingston University.

Archer, H. and Croswell, P. (1989) Public access to GIS: an emerging legal issue. *Photogrammetric Engineering and Remote Sensing*, **55(11)**, 1575–1581.

Arlinghaus, S. L. (1991) Street geometry and flows. *The Geographical Review*, **81**, 206–214.

Aronoff, S. (1989) *Geographic Information Systems: A Management Perspective*, WDL Publications, Canada.

Asrar, G. and Greenstone, R. (eds) (1995) *Mission to Planet Earth: Earth Observation System Reference Handbook*, NASA/Goddard Space Flight Center, USA.

Automobile Association (1986) *3 Mile Road Atlas of Britain*. Automobile Association, Basingstoke.

Ball, D. and Babbage, R. (eds) (1989) *Geographic Information Systems: Defence Applications*, Pergamon Press, Australia.

Ballard, B. (1993) Augusta takes GIS shortcut. *GIS World*, **6**, 20–22.

Banks, D. and Steele, S. (1995) Information resources on the net. *Engineering Science and Education Journal*, **4(3)**, 123–130.

Barnett, D.J. (1997) *Net Research: Finding Information Online*, Songline Studios, New York.

Batty, M. and Xie, Y. (1994) Urban analysis in a GIS environment: population density modelling using ARC/INFO, in *Spatial Analysis and GIS* (eds S. Fotheringham and R. Rogerson), Taylor and Francis, London.

Baxter, R. (1980) A computer file and data structure for manipulating digital outlines of polygons. *Geo Processing*, **1(3)**, 243–256.

Beaumont, J. R. (1991) *An Introduction to Market Analysis*, CATMOG 53, GeoBooks, Norwich.

Bedard, Y. (1987) Uncertainties in land information systems databases. *Proceedings Auto-Carto*, **8**, 175–184.

Benjamin, B. (1968) *Demographic Analysis*, George Allen, London.

Benjamin, B. (1970) *The Population Census*, Heinemann, London.

Benjamin, B. (1989) *Population Statistics – A Review of UK Sources*, Gower, London.

Berhardsen, T. (1992) *Geographic Information Systems*, VIAK IT, Norway.

Bernard, M. (1993) Maintaining schools using FM/GIS: The Languedoc-Roussillon experience. *Geographical Information Systems*, **3(3)**, 46–49.

Black, T. (1992) Local government's one voice. *Mapping Awareness and GIS in Europe*, **6(5)**, 29–30, 47.

Black, T. (1994) A challenge for local government: can the potential of GIS be realised?, in *The 1994 European GIS Yearbook* (eds P.J. Shand and P.J. Ireland), NCC Blackwell, London.

Blakemore, M. (1984) Generalisation error in spatial databases. *Cartographica*, **21**, 131–139.

Blakemore, M. (1990) NOMIS: dynamic local labour market information. *Mapping Awareness*, **4(10)**, 26–30.

Blogg, M. (1990) GIS and the 1991 Census. *Mapping Awareness*, **4(10)**, 10–12.

Board, Chr. (1990) Report of the working group on cartographic definitions. *Cartographical Journal*, **29**, 65–69.

Borden, A. V. (1994) GIS supports election management. *GIS World*, **7(9)**, 24–27.

Briscoe-Taylor, M. (1993) Using GIS to manage assets and risks in local government. *GIS Europe*, **2(1)**, 17–18.

Bromley, D. F. and Selman, J. M. (1993) Geographic information usage in a local authority: the human environment for GIS. *Area*, **25(3)**, 228–236.

Brynum, C. and Wyvill, M. (1992) Detecting changes in urban landscape. *Geographical Information Systems*, **2(6)**, 28–29.

Bulmer, M. (1979) *Censuses, Surveys and Privacy*, Macmillan, London.

Bunce, R. G. H., Morrell, S. K. and Stel, H. E. (1975) The application of multivariate analysis to regional survey. *Journal of Environmental Management*, **3**, 151–165.

Bunce, R. G. H. and Smith, R. S. (1978) *An Ecological Survey of Cumbria*, Cumbria County Council and Lake District Special Planning Board, Kendal, Cumbria.

Bunce, R. G. H., Barr, C. J., Clarke, R. T., Howard, D. C. and Lane, A. M. J. (1996) The ITE Merlewood land classification. *Journal of Biogeography*, **23**, 625–634.

Bundock, M. S. (1987) An integrated DBMS approach to geographical information systems, *Proceedings Auto Carto*, **8**, 292–301.

Burrough, P. A. (1980) The development of a landscape information system in the Netherlands, based on a turn key graphics system. *GeoProcessing*, **1(3)**, 257–274.

Burrough, P. A. (1986) *Principles of Geographical Information Systems for Land Resource Assessment*, Clarendon Press, Oxford.

Burrough, P. A. (1992) The development of intelligent geographical information systems. *International Journal of Geographical Information Systems*, **6(1)**, 1–11.

Burrough, P. A. and McDowell, R. A. (1998) *Principles of Geographical Information Systems*, Oxford University Press, Oxford.

Butland, G. J. (1960) *Latin America: a Regional Geography*, Longman, London.

CACI Information Services (1996) *The Geodemographic Pocket Book*, NTC Publications Ltd, London.

Cadoux-Hudson, J. (1992) *Geographic Information 1992/93*, Taylor and Francis, London.

Callaghan, J. G. (1989) *GIS: Increasing Business Efficiency and Profit*, Proceedings of the Association for Geographic Information Conference, London.

Campbell, H. (1994) How effective are GIS in practice? Case study of British local government. *International Journal of Geographical Information Systems*, **8(3)**, 309–325.

Campbell, H. and Masser, I. (1992) GIS in local government: some findings from Great Britain. *International Journal of Geographical Information Systems*, **6(6)**, 529–546.

Campbell, J. B. (1987) *Introduction to Remote Sensing*, The Guilford Press, New York.

Campbell, W. G. and Masser, I. (1992) The impact of GIS on local government in Great Britain, in *Geographical Information Systems and Urban and Rural Planning* (ed. T. W. Rideout), IBG, Planning and Environment Study Group.

Campbell, W. G., Robbins Church, M., Bishop, G. D., Mortenson, D. C. and Pierson, S. M. (1989) The role of GIS in a large environmental project. *International Journal of Geographical Information Systems*, **3(4)**, 349–362.

Carver, S. Kingston, R. and Turton, I. (1998) *Accessing GIS over the Web: an Aid to Public Participation in Environmental Decision Making*, http://www.geog.leeds.ac.uk/research/papers/98-3

Cassettari, S. (1993) *Introduction to Integrated Geo-information Management*, Chapman and Hall, London.

Castle, G. H. (1993) *Profiting from GIS*, GIS World Inc.

Chambers, J. (1994) Maintaining Grampian's roads: GIS provides a cost-effective route. *GIS Europe*, **3(8)**, 40.

Chatto, C. (1997) Data protection and intellectual property: information systems and the Americanisation of the New Europe. *Environment and Planning A*, **28**, 891–908.

Chorley, R. J. (1988) Reflections on the handling of GIS. *International Journal of Geographical Information Systems*, **2(1)**, 3–9.

Chorley, R. J. and Buxton, R. (1991) The government setting of GIS in the United Kingdom, in *Geographical Information Systems*, Vol I: Principles (eds D. J. Maguire, M. F. Goodchild and D. W. Rhind), Longman Scientific and Technical, Harlow.

Chrisman, N. R. (1983) A theory of cartographic error and its measurement in digital data bases. *Proceedings Auto-Carto*, **5**, 159–168.

Chrisman, N. R. (1987) The accuracy of map overlays: a reassessment. *Landscape and Urban Planning*, **14**, 427–439.

Chuvieco, E. (1990) *Introduction to Teledetection*, Ediciones Rialp, Madrid.

Clark, G. (1982) The Agricultural Census of the United Kingdom and United States. CATMOG, **35**, GeoBooks, Norwich.

Clark, G. (1992) Data sources for studying agriculture, in *The Geography of Agriculture in Developed Market Economies* (ed. I. R. Bowler), London, Longman.

Clarke, J. I., Denham, C., Dewdney, J. C., Evans, I. S., Rhind, D. W. and Visvalingham, M. (1980) *People in Britain: a Census Atlas*, HMSO, London.

Coburn, A. (1993) Which GIS? CICA survey of local authority of GIS usage. *GIS Europe*, **2(7)**, 43.

Cockhead, P. (1985) Population estimates, in *Information Systems for Policy Planning Local Government* (eds J. R. England, K. I. Hudson, R. J. Masters, K. S. Powell and J. D. Shortridge), Longman, Harlow.

Colwell, R. N. (ed.) (1983) *Manual of Remote Sensing*, 2 Vols, American Society for Photogrammetry and Remote Sensing, Virginia, USA.

Community Council for Suffolk (1980) *The Future of Suffolk Villages*, Suffolk County Council, Ipswich.

Cook, B. G. (1978) The structural and algorithmic basis of a geographic database, in *First International Advanced Study Symposium on Topological Data Structures for Geographic Information Systems*, Vol. 4 (ed. G. Dutton), Laboratory for Computer Graphics and Spatial, Harvard University.

Coombes, M. G., Dixon, J. S., Goddard, J. B., Openshaw, S. and Taylor, P. J. (1982) Functional regions for the Population Census of Great Britain, in *Geography and the Urban Environment: Progress in Research and Applications*, Vol V (eds D. T. Herbert and R. J. Johnston), John Wiley and Sons, Chichester.

Coppock, J. (1960) The parish as a geographical statistical unit. *Tijdschrift voor Economische en Sociale Geografie*, **51**, 317–326.

Cousins, S. H., Green, D. R. and Haynes-Young, R. (eds) (1993) *Landscape Ecology and GIS*, Taylor and Francis, London.

Cowan, D. J. and Shirley, P. L. (1993) Integrated planning information system, in *Geographical Information Systems*, Vol II: Applications (eds D. J. Maguire, M. F. Goodchild and D. W. Rhind), Longman Scientific and Technical, Harlow.

Cracknell, A. P. and Hayes, L. W. B. (1993) *Introduction to Remote Sensing*. Taylor and Francis, London.

Crick, H. Q. P., Baillie, S. R., Balmer, D. E., Bashford, R. I., Beaven, L. P., Dudley, C., Glue, D. E., Gregory, R. D., Marchant, J. H., Peach, W. J. and Wilson, A. M. (1998) *Breeding Birds in the Wider Countryside: their Conservation Status (1972–1996)*. BTO Research Report 198, British Trust for Ornithology, Thetford, UK.

Curran, P. J. (1981) *Remote Sensing: the Role of Small Format Light Aircraft Photography*, Geog. Papers, No. 75, Dept. of Geography, University of Reading.

Curran, P. J. (1985) *Principles of Remote Sensing*, Longman, New York.

Curtis, S. E. and Taket, A. R. (1989) The development of geographical information systems for locality planning in health care. *Area*, **21(4)**, 391–399.

Dale, A. (1993) The content of the 1991 Census: change continuity, in *The 1991 Census User's Guide* (eds A. Dale and C. Marsh), HMSO, London.

Dale, A. and Marsh, C. (1993) *The 1991 Census User's Guide*, HMSO, London.

Dale, A., Arber, S. and Procter, M. (1988) *Doing Secondary Analysis*, Allen and Unwin, London.

Data Protection Registrar (1992) *Data Protection Act: the Definitions*, Office of the Data Protection Registrar, Wilmslow.

Data Protection Registrar (1992) *Introduction to the Act*, Office of the Data Protection Registrar, Wilmslow.

Date, S. S. (1991) Emulating the Prospector Expert System with a raster GIS. *Computers and Geosciences*, **17(7)**, 1033–1050.

Davies, S. (1997) Protect us from the snoopers. *New Scientist*, **155**, 48.

Denham, C. (1993) Census geography: an overview, in *The 1991 Census User's Guide* (eds A. Dale and C. Marsh), HMSO, London.

Dent, B. D. (1993) *Cartography: Thematic Map Design*. Wm. C. Brown, Dubuque, IA.

DETR (1998) *Vacant and Derelict Sites and Other Previously Developed Land and Buildings that may be available for Redevelopment: Data Specification*. Department of the Environment, Transport and the Regions, London.

DETR (2000a) *Development Control Statistics, 1999/2000*. Planning and Land Use Statistics Division, Department of the Environment, Transport and the Regions, London.

DETR (2000b) *National Land Use Database Phase I: Supplementary Table Updated Results for Selected Local Authorities in England*. Planning and Land Use Statistics Division, Department of the Environment, Transport and the Regions, London.

DETR (2001) *Land Use Change in England*, No. 15. Department for the Environment, Transport and Regions, London.

Department of the Environment (1990) *Use Classes Order*, HMSO, London.

Dewdney, J. (1981) *The British Census*, Headley Brothers Ltd.

Dickinson, H.J. and Wilkins, H. W. (1988) The economic evolution of implementing a GIS. *International Journal of Geographical Information Systems*, **2(3)**, 307–327.

Dorling, D. (1993) From computer cartography to spatial visualisation: a new cartogram algorithm, *Auto Carto*, **11**, 208–217.

Dorling, D. (1995a) *A New Social Atlas of Britain*, John Wiley and Sons, Chichester.

Dorling, D. (1995b) Visualising changing social structure from a census. *Environment and Planning A*, **26(3)**, 353–378.

Dynes, M. (1990) Bill opens way to reduce roadwork. *The Times*, 8 November 1990, 9.

Eastman, J. R. (1987) *IDRISI: a Grid Based Geographic Analysis System*. Clark University Graduate School of Geography, Worcester, MA.

Edwards, B. (1974) *Sources of Social Statistics*, Heinemann, London.

Ehlers, M., Edwards, G. and Bedward, Y. (1989) Integration of remote sensing with geographic information systems: a necessary evolution. *Photogrammetric Engineering and Remote Sensing*, **55(11)**, 1619–1627.

Ehrensaft, P., LaRamée, P., Bollman, R. D. and Buttel, F. H. (1984) The microdynamics of farm structural change in North America: the Canadian experience and Canada–USA comparisons. *American Journal of Agricultural Economics*, **66(5)**, 823–828.

Elbra, R. A. (1984) *Guide to the Data Protection Act*, National Consumer Council.

Elbra, R. A. (1990) *Data Protection Codes of Practice*, National Consumer Council.

Elkins, P. (1990) Water under the bridge – the background to Southern Water's experience with GIS. *Mapping Awareness*, **4(8)**, 38–40.

England, J. R., Hudson, K. I., Masters, R.J., Powell, K. S. and Shortridge, J. D. (1985) *Information Systems for Policy Planning in Local Government*, Longman, Harlow.

EUROSTAT (1986) *Regions: Statistical Yearbook*, Luxembourg.

EUROSTAT (1987) *REGIO: Regional Data Bank*, EUROSTAT, Division B/2, Luxembourg.

Evans, N.J. (1996) Evaluating recent changes to the publication of UK Agricultural Census data. *Geography*, **352**, 225–234.

Fitzgerald, B. P. (1974) *Developments in Geographical Method*, Oxford University Press, Oxford.

Flowerdew, R. and Martin, D. (1997) *Methods in Human Geography*, Longman, Harlow.

Foresman, T. W. (1992) An honest decade's work pays off. *GIS World*, **5**, 28–30.

Foresman, T. W. (1993) Pueblo County takes the right turn for public safety. *GIS Europe*, **6**, 16–18.

Foresman, T. W. (1994) Tomorrow's promise today's reality. *GIS World*, **7(1)**, 20–24.

Fraser-Taylor, D. R. (1991) *Geographic Information Systems: the Microcomputer and Modern Cartography*, Pergamon Press, Oxford.

General Register Office for Scotland (1998) *Mid–1997 Population Estimates, Scotland*, Edinburgh, General Register Office for Scotland.

Gilster, P. (1994) *The Internet Navigator*, John Wiley, Canada.

Gittus, E. (1963–64) An experiment in the identification of urban sub-areas. *Transactions of the Bartlett Society*, **2**, 108–135.

Goodchild, M. F. (1978) *Statistical Aspects of the Polygon Overlay Problem*, Harvard Papers on Geographic Information Systems, 6. Addison-Wesley, Reading, MA.

Goodchild, M. F. (1988) The issue of accuracy in global databases, in *Building Databases for Global Science* (eds H. Mounsey and R. Tomlinson), Taylor and Francis, London, pp. 31–48.

Goodchild, M. F. (1992) Geographical data modelling: the raster vector debate. *Computers in Geoscience*, **18(4)**, 401–408.

Goodchild, M. F. (1995) GIS and geographic research, in *Ground Truth: The Social Implications of Geographic Information Systems* (ed. J. Pickles), Guilford Press, New York.

Goodchild, M. F. and Gopal, S. (1989) *Accuracy of Spatial Databases*, Taylor and Francis, London.

Goodwin, T. (1997) Fighting the supermarket sweep. *Mapping Awareness*, **11(5)**, 16–18.

Gordon, A. and Wormsley, J. (1997) The use of mapping in public health and planning health services. *Journal of Public Health Medicince*, **19(2)**, 139–147.

Goss, J. (1995) We know who you are and we know where you live: the instrumental rationality of geodemographic systems. *Economic Geography*, **17(2)**, 171–174.

Gould, P. and White, R. (1974) *Mental Maps*, Penguin, Harmondsworth.

Green, D. R., Rix, D. and Cadoux-Hudson, J. (1993) *Geographic Information 1994: the Source Book for GIS*, Taylor and Francis, London.

Grimshaw, D.J. (1989) GIS: a tool for business and industry. *International Journal of Information Management*, **9**, 119–126.

Grimshaw, D.J. (1995) *Bringing Geographical Information Systems into Business*, GeoInformation International, Cambridge.

Hakim, C. (1982) *Secondary Analysis in Social Research*, Allen and Unwin, London.

Hammersley, C. (1993) *Social Research: Philosophy, Politics Practice*, Sage Publications, London.

Hammond Incorporated (1996) *Concise Atlas of the World*, Oxford University Press, Oxford.

Handly, M. and Crowcroft, J. (1995) *The World Wide Web: Beneath the Surf*, UCL Press, London.

Hannavy, M. (1996) *Introducing the Data Archive*, BIRON_Internet, University of Essex, Colchester.

Haralick, R. M. (1980) A spatial data structure, in *Map Data Processors* (eds H. Freeman and G. G. Pieroni), Taylor and Francis, London.

Harder, C. (1998) *Serving Maps on the Internet: Geographic Information on the World Wide Web*, ESRI, USA.

Hargreaves, P. and Spedding, V. (1995) Materials on line: information from the net. *Materials World*, **3(10)**, 483–484.

Harris, R. (1988) *Satellite Remote Sensing: An Introduction*, Routledge & Kegan Paul, London and New York.

Harvey, D. (1969) *Explanation in Geography*, Edward Arnold, London.

Healey, M.J. (1991) *Economic Activity and Land Use: the Changing Information Base for Local Regional Studies*, Longman, Harlow.

Healey, R. (1988) Interfacing software systems for geographical information systems. *ESRC Newsletter*, **63**, 23–26.

Herbert, J. (1984) *Using Published Data: Errors and Remedies*, Sage Publications Inc, London.

Heywood, I., Cornelius, S. and Carver, S. (1998) *An Introduction to Geographical Information Systems*, Longman, Harlow.

Hill, M. O. (1979a) *DECORANA – a FORTRAN Program for Detrended Correspondence Analysis and Reciprocal Averaging*. Section of Ecology and Systematics, Cornell University, Ithaca, NY.

Hill, M. O. (1979b) *TWINSPAN – a FORTRAN Program for Arranging Multivariate Data in an Ordered Two-way Table by Classification of the Individuals and Attributes*, Section of Ecology and Systematics, Cornell University, Ithaca, NY.

HMSO (1987) *Handling Geographic Information*, Report of the Chorley Committee of Enquiry.

HMSO (1991) *1991 Census of Population – Confidentiality and Computing*, Cmnd. 1447, HMSO, London.

HMSO (1993a) *Geographic Information Systems: A Buyers Guide*, Government Centre for Information Systems, London.

HMSO (1993b) *Geographic Information Systems: Reaching the Opportunities*, Government Centre for Information Systems, London.

HMSO (1994) *A Survey of Geographic Information Systems in Government*, Government Centre for Information Systems, London.

HMSO (2001) *Cabinet Office News Release*, http://www.hmso.goc.uk/docs/mccartney.htm

Hobcraft, J. Marsh, C., Lievesley, D., Penhale, B., Openshaw, S., Skinner, C. and Walford, N. S. (1991) The case for samples of anonymised records from the 1991 Population Census. *Journal of the Royal Statistical Society*, **154(2)**, 305–340.

Hooker, P.J., Ellison, R.A. Marchant, A.P., Shaw, R.P., Leader, R.U., Newsham, R. and Brown, C.M.J. (2000) *Some Guidance on the Use of Digital Environmental Data*, British Geological Survey Report WE/99/14, Environment Agency Report NC/06/32.

Hughes, D. (1988) The basics of a Geographic Information System. *Mapping Awareness*, **1(10)**, 10–12.

Ibbs, T.J. and Stevens, A. (1988) Quadtree storage of vector data. *International Journal of Geographical Information Systems*, **2(1)**, 43–56.

Ireland, P. (1994) Local government GIS survey trends reveal all. *Mapping Awareness*, **6(6)**, 14–15.

Irvine, J. and Miller, I. (1979) *Demystifying Social Statistics*, Pluto Press Ltd, London.

James, F.C. and Shugart, H.H., Jr (1970) A quantitative method of habitat description. *Audubon Field Notes*, **24**, 727–736.

James, S. and Pope, R. (1993) Corporate GIS in local government: the Horsham experience. *Mapping Awareness and GIS Europe*, **7(4)**, 28–31.

Janner, G. (1984) *Data Protection*, Croner Publications Ltd, New Malden.

Johnston, D.W. (1990) Descriptions of surveys: Breeding Bird Census, in *Survey Designs and Statistical Methods for the Estimation of Avian Population Trends* (eds J.R. Sauer and S. Droege), US Fish and Wildlife Services Biological Report, No. 90(1).

Johnston, D.W. and Hagan, J.M., III (1992) An analysis of long-term Breeding Bird Censuses from eastern deciduous forests, in *Ecology and Conservation of Neotropical Migrant Landbirds* (eds J.M. Hagan III and D.W. Johnston), Smithsonian Institute Press, Washington DC.

Jones, A. (1996) Using digital imagery: Part 2 remote sensing. *Mapping Awareness*, **10(7)**, 24–26.

Jones, C.B. (1997) *Geographical Information Systems and Computer Cartography*, Longman, Harlow.

Juhl, G.M. (1994) City relies on GIS to reduce costs, increase productivity. *GIS World*, **7(5)**, 43–45.

Kaiser, J. (1997) Census plans more sampling. *Science*, **275**, 1727.

Kennedy, A.J. (1995) *The Internet and the World Wide Web*, Penguin, London.

Kerfoot, G. (1993) Regional mall site selection. *Geographical Information Systems*, **3(6)**, 51–54.

Kiernan, V. (1994) How to make an imperfect census count. *New Scientist*, **26**, 9.

Kinch-James, D. (1993) Where and why? – Market analysis, the key to business success. *Mapping Awareness*, **7(7)**, 2–7.

Kingdom, B. (1991) The use of a GIS for rehabilitating planning within the water industry. *Mapping Awareness*, **5(1)**, 9–11.

Knaap, van der W.G.M. (1992) The vector to raster conversion: (Mis)use in geographical information systems. *International Journal of Geographical Information Systems*, **6(2)**, 159–170.

Knott, J. and Shiers, D. (1994) Beam me up a map, Scotty! *Mapping Awareness*, **8(2)**, 32–34.

Kraak, M.J. (1993) A GIS for Dutch river management. *GIS Europe*, 30–32.

Kraak, M.J. and Ormeling, F.J. (1996) *Cartography: Visualisation of Spatial Data*, Longman, Harlow.

Kramer, H.J. (1994) *Observation of the Earth and its Environment: Survey of Missions and Sensors*, Springer-Verlag, Berlin, New York and London.

Laming, R. (1993) GIS supporting our services. *Mapping Awareness and GIS Europe*, **7(4)**, 11–14.

Land Information New Zealand (2000) *NZTopo Data Documentation Guide: Topographic Data for GIS, Primary Source Scale 1 : 50,000*, National Topographic/Hydrographic Authority.

Langran, G. and Chrisman, N. R. (1988) A framework for temporal geographic information. *Cartographica*, **25(3)**, 1–14.

Lews, F. M. M. and Parker, S. R. (1975) *Reviews of UK Statistical Sources: Vols II and IV*, Heinemann Education Books, London.

Lillesand T. M. and Kiefer, R. W. (2000) *Remote Sensing and Image Interpretation*, John Wiley and Sons, New York.

Lloyd, C. (1994) Area you ready for the future? *The Sunday Times*, November 20, 16.

Lodgwick, A. and Cushnie, J. A. (1990) A GIS pilot study in Berkshire. *Mapping Awareness*, **4(10)**, 39–42.

Longley, P. and Clarke, G. (1995) *GIS for Business and Service Planning*, GeoInformation International, Cambridge.

Lopez, X. R. and John, S. A. (1993) Data exchange and integration in the UK local government. *Mapping Awareness and GIS Europe*, **7(5)**, 37–40.

Lunn, D.J., Simpson, S. N., Diamond, I. and Middleton, L. (1998) The accuracy of age-specific population estimates for small areas in Britain. *Population Studies*, **52**, 327–344.

Lyon, F. (1989) Implementing a GIS: the Hampshire experience. *Mapping Awareness*, **3(6)**, 19–21.

Madwick, D. and Smythe, T. (1974) *The Invasion of Privacy*, Sir Isaac Pitman & Sons Ltd, London.

Maffinni, G. (1987) Raster versus vector data encoding and handling: a commentary. *Photogrammetric Engineering and Remote Sensing*, **53(10)**, 1397–1398.

Maguire, D.J. (1989) *Computers in Geography*, Longman Scientific and Technical, Harlow.

Maguire, D.J. (1991a) Implementing GIS: local authorities learn how to plan for success. *Mapping Awareness*, **5(8)**, 3.

Maguire, D.J. (1991b) The CAMP project – meeting national needs. *Mapping Awareness*, **5(8)**, 6.

Maguire, D.J. (1992) The raster GIS design model: a profile of ERDAS. *Computers and Geosciences*, **18(4)**, 463–470.

Maguire, D.J. and Raper, J. F. (1992) Design models and functionality in GIS. *Computers and Geosciences*, **18(4)**, 387–394.

Maguire, D.J., Goodchild, M. F. and Rhind, D. W. (eds) (1993a) *Geographical Information Systems: Principles*, Vol. I, Longman Scientific and Technical, Harlow.

Maguire, D.J., Goodchild, M. F. and Rhind, D. W. (eds) (1993b) *Geographical Information Systems: Applications*, Vol. II, Longman Scientific and Technical, Harlow.

Mahon, B. (1995) Census confidentiality. *Mapping Awareness*, **9(7)**, 40–41.

Marchant, J. H. (1983) *BTO Common Birds Census Instructions*, British Trust for Ornithology, Tring, UK.

Mark, D. M. (1986) The use of quadtrees in geographic information systems and spatial data handling. *Proceedings of Auto Carto*, **1**, 517–526.

Marsh, C. (1993) Privacy, confidentiality and anonymity in the 1991 Census, in *The 1991 Census User's Guide* (eds A. Dale and C. Marsh), HMSO, London.

Marsh, C., Dale, A. and Skinner, C. (1993) Safe data versus safe settings: access to microdata from the British Census. *International Statistical Review*, **62(1)**, 35–53.

Martin, D. (1993) *The UK Census of Population 1991*, Environmental Publications, Norwich.

Martin, D. (1992) Postcodes and the 1991 Census of Population: issues, problems and prospects. *Transactions of the Institute of British Geographers*, **17**, 350–357.

Martin, D. (1996) *Geographic Information Systems and their Socio-Economic Applications*, 2nd edn, Routledge, London.

Martin, D. (1997) From enumeration districts to output areas: experiments in the automated creation of a census output geography. *Population Trends*, **88**, 36–42.

Martin, D. (2000) Towards the geographies of the 2001 UK Census of Population. *Transactions of the Institute of British Geographers*, **25(3)**, 321–332.

Martin, D. and Bracken, I. (1991) Techniques for modelling population related raster databases. *Environment and Planning A*, **23(7)**, 1069–1075.

Masser, I. and Blakemore, M. (1991) *Handling Geographic Information*, Longman, Harlow.

Massey, D. (1999) Space-time, 'science' and the relationship between physical and human geography. *Transactions of the Institute of British Geographers*, **24(3)**, 261–276.

Mather, P.M. (1987) *Computer Processing of Remotely Sensed Images*, John Wiley and Sons, Chichester.

Mather, P.M. (1994a) *Computer Applications in Geography*, John Wiley and Sons, Chichester.

Mather, P.M. (1994b) *Geographical Information Handling – Research and Applications*, John Wiley and Sons, Chichester.

Matthews, H. and Foster, I. (1989) *Geographical Data: Sources, Presentation and Analysis*, Oxford University Press, Oxford.

McKinsey, D., Mertz, Nagel, D., Nathanson, J., Spery, S., Stow, D. and Westmoreland, S. (1990) Efficient creation, correction and updating of vector-coded GIS coverage using remotely sensed data. *GIS/LIS '90 Proceedings*, Vol. 1, 209–218.

Middleton, L., Simpson, S.N., Diamond, I. and Lunn, D. (1997) *An Experiment with Methods of Estimating the Population of Small Areas in Britain*, Estimating with Confidence Project, Working Paper No. 14. University of Southampton, Southampton.

Miller, P. (1994) Medical centre uses desktop mapping to cut costs and improve efficiency. *Geographical Information Systems*, **4(4)**, 40–44.

Mitchell, G. (1995) Mapping the landscape for Belfast City Council. *Mapping Awareness*, **9(1)**, 24.

Moorhouse, S. (1985) ARC/INFO: a geo-relational model for spatial. *Proceedings Auto Carto London*, **7**, 420–427.

Morphet, C. (1992) The interpretation of small area census data. *Area*, **24(1)**, 63–72.

Mounsey, H. (1991) Multisource, multinational GIS: lessons learnt from CORINE, in *Geographical Information Systems: Principles and Applications*, Vol 2 (eds D.J. Maguire, M.F. Goodchild and D.W. Rhind), Longman, London, 185–200.

Myers, D. (1992) *Analysis of Local Census Data: Portraits of Change*, Academic Press, London.

Nagel, D. and Robinson, R. (1990) Land cover classification of remotely-sensed imagery and conversion to a vector based GIS for the Swannee River Water Management District. *GIS/LIS '90 Proceedings*, Vol. 1, 219–224.

Naisbitt J. (1991) *Megatrends 2000: the Next Ten Years – Major Changes in your Life and World*, Pan (Sidgwick and Jackson), London.

National Academy of Sciences (1995) *A Data Foundation for the National Spatial Data Infrastructure*, National Academy Press, Washington DC.

National Research Council (1993) *Towards a Co-ordinated Spatial Data Infrastructure*, National Academy Press, Washington, DC.

Norton-Taylor, R. (1995) Free in statutes for the net. *The Guardian*, October 18, 2.

O'Looney, J. (2000) *Beyond Maps: GIS and Decision making in Local Government*, ESRI, Redlands, CA.

Office for National Statistics (1997) *The ONS Classification of Local and Health Authorities in Great Britain*, HMSO, London.

Office for National Statistics (1998) *Key Population and Vital Statistics: Local and Health Authority Areas*, HMSO, London.

Office for National Statistics (2000) *Living in Britain: results from the General Household Survey*, HMSO, London.

Office for National Statistics/General Register Office (Scotland)/Northern Ireland Statistical and Research Agency (1999) *2001 Census: a Guide to the One Number Census*, http://www.ons.gov.uk/census/onc.pdf

Openshaw, S. (1977) A geographical solution to scale aggregation problems in region-building, partitioning and spatial modelling. *Transactions of the Institute of British Geographers*, **2(4)**, 459–472.

Openshaw, S. (1984) *The Modifiable Areal Unit Problem*, CATMOG, No. 38. Geo Abstracts, Norwich.

Openshaw, S. (1987) A mark 1 geographical information analysis machine for the analysis of point datasets. *International Journal of Geographical Information Systems*, **1**, 335–338.

Openshaw, S. (1989) Errors in spatial databases, in *Accuracy of Spatial Databases* (eds M. Goodchild and S. Gopal), Taylor and Francis, London.

Openshaw, S. (ed.) (1995) *Census Users' Handbook*, GeoInformation International, Cambridge.

Openshaw, S., Wymer, C. and Charlton, M. (1986) A geographical information and mapping system for the BBC Domesday optical disks. *Transaction of the Institute of British Geographers*, **11(3)**, 296–304.

Parent, P. (1991) Co-operation a key to systems success. *GIS World*, **4**, 28.

Parrott, R. and Stutz, F. P. (1993) Urban GIS applications, in *Geographical Information Systems: Vol II: Applications* (eds D.J. Maguire, M. F. Goodchild and D. W. Rhind), Longman Scientific and Technical, Harlow.

Peel, R. (1993) Departmental GIS pilots: a catalyst for data integration. *Mapping Awareness and GIS Europe*, **7(6)**, 38–40.

Peel, R. (1995) Fast track to GIS PC-based systems in local government. *Mapping Awareness and GIS Europe*, **9(3)**, 30–33.

Peers, E. and Buckley, B. (1988) *Computers and Data Protection*, Deloitte, Haskins and Sells, London.

Petersen, W. (1975) *Population*, 3rd edn, Macmillan, London.

Peucker, T. K. and Chrisman, N. (1975) Cartographic data structures. *American Cartography*, **2(1)**, 55–69.

Peuquet, D.J. and Marble, D. F. (1990) *Introductory Readings in GIS*, Taylor and Francis, London.

Pickett, K. C. (1974) *Sources of Official Data*, Longman Group, London.

Pinpoint Analysis (1990) Pinpoint: unlocking the value of your data. *Mapping Awareness*, **4(9)**, 3–6.

Plane, D. A. and Rogerston, A. (1994) *The Geographical Analysis of Population with Applications to Planning Business*, John Wiley and Sons, Chichester.

Rabbitt, M. C. (1989) *The United States Geological Survey, 1879–1989*, USGS Circular 1050.

Ramage, P. (1994) Information in, rubbish out. *Mapping Awareness*, **8(4)**, 34–36.

Raper, J. and Livingstone, D. (1995) Development of a geomorphological spatial model using object-oriented design. *International Journal of Geographical Information Systems*, **9**, 359–383.

Raper, J. and Maguire, D.J. (1992) Design models and functionality in GIS. *Computers in Geoscience*, **18(4)**, 387–394.

Raper, J., Rhind, D.W. and Shepherd, J.W. (1993) *Postcodes the New Geography*, Longman, Harlow.

Rees, P.H. (1998) What do you want from the 2001 Census? Results of an ESRC/JISC survey of user views. *Environment and Planning A*, **30**, 1775–1796.

Rees, P.H. and Duke-Williams, O. (1998) Can Census Offices publish statistics for more than one small area geography? An analysis of the differencing problem in statistical disclosure. *International Journal of Geographical Information Science*, **12(6)**, 579–605.

Rees. P.H. (ed.) (1997*) The Debate about the Geography of the 2001 Census: Collected Papers from 1995–96*, WP97/01. School of Geography, University of Leeds, Leeds.

Rees, P.H. and Martin, D. (1997) Flexible geographies and area aggregation: designing small areas for outputs from the 2001 Census, in *The Debate about the Geography of the 2001 Census: Collected Papers from 1995–6* (ed. P.H. Rees), Working Paper 97/1, School of Geography, University of Leeds.

Rhind, D.W. (1977) Computer-aided cartography. *Transactions of the Institute of British Geographers*, **2**, 71–96.

Rhind, D.W. (1981) GIS in the UK, in *Quantitative Geography* (eds N. Wrigley and R.J. Bennett), Routledge, Kegan and Paul, London.

Rhind, D.W. (1983) *A Census User's Handbook*, Methuen, London.

Rhind, D.W. (1987) Recent developments in GIS. *International Journal of Geographical Information Systems*, **1(3)**, 229–241.

Rhind, D.W. (1988) Computing, academic geography and the outside world, in *Remodelling Geography* (ed. W. MacMillan), Blackwell, Oxford.

Rhind, D.W., Wyatt, B.K., Briggs, D.J. and Wiggins, J.C. (1986) The creation of an environmental information system for the European Community. *Nachrichter aus dem Karten und Veressungsvesen Series 2*, **44**, 147–157.

Robbins, C.S. (1970) Recommendations for an international standard for a mapping method in bird census work. *Audubon Field Notes*, **24**, 723–726.

Robinson, G.M. (1998) *Methods and Techniques in Human Geography*, John Wiley and Sons, Chichester.

Robson, B.T. (1969) *Urban Analysis: a Study of City Structure with Special Reference to Sunderland*, Cambridge University Press, Cambridge.

Roper, C. (1994) Hydrographic office bids for world market with Admiralty raster chart service. *Mapping Awareness*, **8(5)**, 36–39.

Rourk, R.W. (1993) How good is GIS in local government. *GIS World*, **6**, 53–56.

Rzepa, H.S. (1996) Science and the internet: the WWW. *Science Progress*, **79(2)**, 97–117.

Saeed, S.E.O. (1993) British Telecom uses GIS to gear investment into the marketplace. *GIS Europe*, **2(1)**, 10–12.

Samet, H. (1990) *The Design and Analysis of Spatial Data*, Addison-Wesley, Reading, MA.

Schofield, S. (1995) *The UK Internet Book*, Addison-Wesley Longman, London.

Scottish Office (1980) *Report on a Pilot Project for a Rural Land Use Information System*, RLUIS Working Party. Scottish Office, Edinburgh.

Shand, P.J. (1987) Uses for the scanner's basic raster image. *Mapping Awareness*, **1(8)**, 3–4.

Shapiro, D., Bollman, R.D. and Ehrensaft, P. (1987) Farm size and growth in Canada. *American Journal of Agricultural Economics*, **69(2)**, 479–483.

Shaw, G. and Wheeler, D. (1985) *Statistical Techniques in Geographical Analysis*, David Fulton, London.

Shelton, R. L. and Estes, J. E. (1979) *Integration of Remote Sensing and Geographical Information Systems*, Ann Arbor, University of Michigan, Environmental Research Institute, 675–692.

Short, B. M. and Watkins, C. (1994) The National Farm Survey of England and Wales 1941–43, *Area*, **26**, 288–293.

Short, B. M. and Watkins, C. (1999) Prelude to modernity: an evaluation of the National Farm Survey of England and Wales, 1941–43, in *Reshaping the Countryside: Perceptions and Processes of Rural Change* (eds N. S. Walford, J. Everitt and D. Napton), CABI, Wallingford.

Silk, J. (1979) *Statistical Concepts in Geography*, Allen and Unwin, London.

Simpson, S., Diamond, I., Tonkin. P. and Tye, R. (1996) Updating small area population estimates in England and Wales. *Journal of the Royal Statistical Society, Series A*, **159**, 235–247.

Sizer, R. and Newman, P. (1984) *The Data Protection Act*, Gower, London.

Skelton, B. and Yost, M. (1990) Programming language technology for raster GIS modelling, *GIS/LIS '90 Proceedings*, Vol. 1, 317–327.

Slattery, M. (1986) *Official Statistics*, Tavistock Publications, London.

Sleight, P. (1995) Neighbourhood watch: geodemographic and lifestyle data in the UK marketplace. *Mapping Awareness*, **9(6)**, 18–21.

Smith D. G. and McBride G. B. (1990) New Zealand's national river water quality monitoring network-design and first year's operation. *Water Resources Bulletin*, **26(5)**, 767–775.

Smith D. G., McBride G. B., Bryos, E. G., Wisse, J. and Mink, D. F. J. (1996) Trends in New Zealand's National Rivers Water Quality Network. *New Zealand Journal of Marine and Freshwater Research*, **30**, 485–500.

Smith, T. R., Menon, S., Starr, J. L. and Estes, J. E. (1987) Requirements and principles for large scale GIS. *International Journal of Geographical Information Systems*, **1(1)**, 13–31.

Smith, G. R. and Honeycutt, D. M. (1987) *Geographic Data Uncertainty, Decision Making and the Value of Information*, Proceedings, Second Annual International Conference, Exhibits and Workshops on Geographic Information Systems: 300–312.

Smith, N. (1988) Data structures. *Mapping Awareness*, **1(10)**, 7–8.

Somers, R. and Singh, Y. (1992) Developing an accurate database using GPS. *GIS World*, **5**, 22–24.

Sonka, M., Hlavac, V. and Boyle, R. (1993) *Image Processing, Analysis and Machine Vision*, Chapman and Hall, London.

Speed, V. (1994) County employs GIS for integrated emergency management. *GIS World*, **7(4)**, 28–33.

Star, E. and Estes, J. E. (1990) *Geographical Information Systems: an Introduction.* Englewood Cliffs, New Jersey.

Stark, B. (1993) Making a mint out of GIS. *Mapping Awareness*, **6(6)**, 30–32.

Stedman, J. R. (1998). *Revised High Resolution Maps of Background Air Pollutant Concentrations in the UK: 1996.* AEA Technology, National Environmental Technology Centre.

Steigler, S. E. (1978) *Dictionary of Earth Sciences*, Pan Books, London.

Stow, D. and Westmoreland, S. J. (1990) Use of satellite imagery and ancillary data to update a vector-coded GIS. *GIS/LIS '90 Proceedings*, Vol. 1, 383–391.

Streich, T. A. (1986) *Geographic Data Processing: a Contemporary Overview*, Masters Thesis, University of California, Dept. of Geography.

Stuart, N. (1990) Quadtree GIS-pragmatics from the present prospects for the future. *GIS/LIS '90 Proceedings*, Vol. 1, 373–382.

Teicholz, E. and Berry, B. J. L. (1983) *Corporate Graphics and Environmental Planning*, Prentice Hall, London.

Thapa, K. (1988) Critical points detection and automatic line generalisation in raster data using zero-crossings. *The Cartographic Journal*, **25**, 58–67.

Thomas, B.J. (1995) *The Internet for Scientists and Engineers*, SPIE Press, Washington.

Thrall, G. I. and Smersh, G. (1994) Business GIS data, Part Four: attribute data. *Geographical Information Systems*, March.

Tomlin, C. D. (1990) *Geographic Information Systems and Cartographic Modelling*, Prentice Hall, Englewood Cliffs.

Tomlinson, D. (1987) Current issues in GIS in North America. *International Journal of Geographical Information Systems*, **1(3)**, 203–218.

Tomlinson, R. F. (1988) The impact of the transition from analogue to digital cartographic representation. *American Cartographer*, **15(3)**, 249–263.

Toon, M. (1998) Get within the script. *Mapping Awareness*, **12(11)**, 14.

Townsend, A., Blakemore, M., Nelson, R. and Dodds, P. (1986) The National On-Line Manpower Information System (NOMIS). *Employment Gazette*, **94**, 60–64.

Townsend, A., Blakemore, M. Nelson, R. and Dodds, P. (1987) The NOMIS database: availability and uses for geographers. *Area*, **19**, 43–50.

Treadwell, J. H. (1995) Behind the scenes: land satellite information in the next decade. *Photogrammetric Engineering and Remote Sensing*, **61**, 989–991.

Tucker, C.J. (1979) Red and photographic infrared linear combinations for monitoring vegetation. *Remote Sensing of Environment*, **10**, 127–150.

Veregin, H. (1989) Error modelling for the map overlay operation, in *Accuracy of Spatial Databases* (eds M. Goodchild and S. Gopal), Taylor and Francis, London.

Wadsworth, R. and Treweek, J. (1999) *Geographical Information Systems for Ecology: an Introduction*, Longman, Harlow.

Walford, N. S. (1995) *Geographical Data Analysis*, John Wiley and Sons, Chichester.

Walford, N. S. (1997) Evolution and development of large-scale farming in South East England, 1941–1981, in *The Kingston Papers: a Geographical Perspective on the Environment, Economy and Society* (eds J. A. Holmes and K. D. Lynch), School of Geography, Kingston University, Kingston upon Thames.

Walford, N. S. and Hockey, A. E. (1991) Sources for the study of social change in the countryside, in *People in the Countryside* (eds T. Champion and C. Watkins), Paul Chapman, London.

Walford, N. S. and Hockey, A. E. (2000) Monitoring demographic and socioeconomic characteristics of intercensal populations: a British local authority perspective. *Planning Practice and Research*, **15(1/2)**, 51–63.

Walford, N. S., Lane, A., Shearman, J. and Newby, H. N. (1989) *Rural Areas Database Catalogue*, ESRC Data Archive, University of Essex, Colchester.

Walford, N. S., Lane, A., Shearman, J. and Newby, H. N. (1989) The Rural Areas Database: a geographical information mapping system for the management integration of data on rural Britain. *Transactions of the Institute of British Geographers*, **16(2)**, 221–231.

Waniez, P. (1994) Software review: forming an alliance. *GIS Europe*, **3(10)**, 44.

Ward, B. E. (1990) User-oriented format for topological vector databases. *GIS/LIS '90 Proceedings*, Vol. 1, 402–412.

Warner, W. S., Graham, R. W. and Read, R. E. (1996) *Small Format Aerial Photography*, American Society for Photogrammetry and Remote Sensing, Maryland.

Waugh, T. C. (1986) *A Response to Recent Papers and Articles on the Use of Quadtrees for Geographic Information*, Proceedings of the Second International Symposium on Spatial Handling, Seattle, Washington.

Waugh, T. C. and Taylor, D. R. F. (1977) GIMMS: an example of an operational system for computer cartography, *Canadian Cartographer*, **13**, 158–166.

Webb, R. and Elkins, P. (1989) Guidelines for the implementation and management of a GIS. *Proceedings of the Association for Geographic Information Conference*: 3.3.1–3.3.6.

Weber, W. (1978) Three types of map data structures, their ANDs and NOTs and a possible OR, in *First International Advanced Study Symposium on Topological Data Structures for Geographic Information Systems* (ed. G. Dutton), Laboratory for Computer Graphics and Spatial Analysis, Harvard University.

Wehde, M. (1982) Grid cell size in relation to errors in maps and inventories produced by computerised mapping process. *Photogrammetric Engineering and Remote Sensing*, **48(8)**, 1289–1298.

Welsh Consumer Council (1980) *What's Where in West Glamorgan?* Welsh Consumer Council, Cardiff.

Whitehead, F. (1988) How the 1991 Census should improve government statistics. *Population Trends*, **53**, 18–21.

Wiley, B. (1994) Tackling the data representation problem using both vector and raster GIS. *Geographical Information Systems*, **4(4)**, 28–29.

Williams, G. (1985) Recent social chances in Mid-Wales. *Cambria*, **12(2)**, 117–138.

Williams, R. B. G. (1985) *Introduction to Statistics for Geographers and Earth Scientists*, Macmillan, London.

Williams, W. T. and Lambert, J. M. (1959) Multivariate methods in plant ecology. I. Association analysis in plant communities. *Journal of Ecology*, **47**, 83–101.

Worboys, M. F. (ed.) (1994) *Innovations in GIS*, Taylor and Francis, London.

Worrall, L. (1990) *Geographic Information Systems: Developments and Applications*, Belhaven Press, Edinburgh.

Worrall, L. (1994) Justifying investment in GIS: a local government perspective. *International Journal of Geographical Information Systems*, **8(6)**, 543–565.

Wrigley, N. (1985) *Categorical Data Analysis for Geographers and Environmental Scientists*, Longman, Harlow.

Youngs, F. A. (1979) *Guide to the Local Administration Units of England*, Offices of the Royal Historical Society, University College London, Gower Street.

Appendix

Household Form for the 2001 UK Census of Population

H1

29 April
count me in
Census2001

England Household Form

Name

Address

Postcode

CD

ED

Form Number

* Form 1 of

*Multi-form households only

SPECIMEN

To the Householder, Joint Householders or members of the household aged 16 or over

The Census is a count every ten years of all people and households in the country. Census information is used by central and local government, health authorities and many other organisations to allocate resources and plan services for everyone. The Office for National Statistics conducts the Census in England and Wales.

Completing your form

Completion of the Census form is compulsory under the Census Act 1920. If you refuse to complete it, or give false information, you may be liable to a fine. This liability does not apply to question 10 on religion. The requirement for you to return a completed form will not be satisfied until such a form has been received. If you need help please contact the Census Helpline.

Confidentiality

The information you provide is protected by law and treated in strict confidence. The information is only used for statistical purposes, and anyone using or disclosing Census information improperly will be liable to prosecution. Census forms will be held securely. Under the current terms of the Public Records Act 1958, the data will be treated as confidential for a period of 100 years.

Thank you for counting yourself in.

Len Cook
REGISTRAR GENERAL FOR ENGLAND AND WALES

national STATISTICS

What you have to do

- Your household should complete this form in **black or blue ink**. A household is:
  - one person living alone, or
  - a group of people (not necessarily related) living at the same address with common housekeeping - sharing either a living room or sitting room, or at least one meal a day.
- This form covers five people. If there are more than five people in your household you will need an extra form.
- Identify household members in Table 1 (page 2). It will help you to complete the form if you use Table 2 to identify visitors.
- Answer the questions about your accommodation (page 3).
- Complete the relationship question (pages 4 and 5).
- Answer the remaining questions for every member of your household.
- Sign the Declaration and **post the form back** in the envelope supplied.

Declaration

- To be signed after completing this form. Please check that you have not missed any pages or questions.

This form is completed to the best of my knowledge and belief.

Signature/s | **Date**

Table 1 Household Members

- **List all members of your household who usually live at this address, including yourself.**
  - **Start with the Householder or Joint Householders.**
  - **Include anyone who is temporarily away from home on the night of 29 April 2001 who usually lives at this address.**
  - **Include schoolchildren and students if they live at this address during the school, college or university term.**
  - **Also include schoolchildren and students who are away from home during the school, college or university term and for whom only basic information is required.**
  - **Include any baby born before 30 April 2001, even if still in hospital.**
  - **Include people with more than one address if they live at this address for the *majority of time*.**
  - **Include anyone who is staying with you who has no other usual address.**
  - **Remember to include a spouse or partner who works away from home, or is a member of the armed forces, and usually lives at *this address*.**
- **If any member of your household aged 16 or over requires a separate form for privacy reasons, please contact the Census Helpline and ✓ the relevant box in the column marked 'Individual Form'.**

Person No.	First name and surname	Individual Form
Person 1		☐
Person 2		☐
Person 3		☐
Person 4		☐
Person 5		☐
If you have more than 5 people in your household, you will need an extra form.		
Person 6		☐
Person 7		☐
Person 8		☐
Person 9		☐
Person 10		☐

Table 2 Visitors

- **To help you complete the form you may use Table 2 to list any visitors at this address, on the night of 29 April 2001, who usually live elsewhere.**
- **If there are only visitors at this address, please complete questions H1 to H5 on page 3. No further questions need to be answered.**

First name and surname	Usual address

How to complete the remaining questions

Remember to use black or blue ink.

Put a tick in the appropriate box, like this ☑. If you mark the wrong box, fill in the box and put a tick in the right one, like this ■ ☑

Where you are required to write in an answer please use CAPITAL LETTERS and leave one space between each word. Start a new line if a word will not fit.

7 What is your country of birth?

☑ Elsewhere, *please write in the present name of the country*

S O U T H

A F R I C A

Household Accommodation

H1 What type of accommodation does your household occupy?

A *whole* house or bungalow that is:

☐ Detached

☐ Semi-detached

☐ Terraced (including end-terrace)

A flat, maisonette, or apartment that is:

☐ In a purpose-built block of flats or tenement

☐ Part of a converted or shared house (includes bed-sits)

☐ In a commercial building (for example, in an office building, or hotel, or over a shop)

Mobile or temporary structure:

☐ A caravan or other mobile or temporary structure

H2 Is your household's accommodation self-contained?

- This means that *all* the rooms, including the kitchen, bathroom and toilet are behind a door that only your household can use.

☐ Yes, all the rooms are behind a door that only our household can use

☐ No

H3 How many rooms do you have for use only by your household?

- *Do not count* bathrooms, toilets, halls or landings, or rooms that can only be used for storage such as cupboards.
- *Do count* all other rooms, for example kitchens, living rooms, bedrooms, utility rooms and studies.
- If two rooms have been converted into one, count them as one room.

Number of rooms ☐☐

H4 Do you have a bath/shower *and* toilet for use only by your household?

☐ Yes

☐ No

H5 What is the lowest floor level of your household's living accommodation?

☐ Basement or semi-basement

☐ Ground floor (street level)

☐ First floor (floor above street level)

☐ Second floor

☐ Third or fourth floor

☐ Fifth floor or higher

H6 Does your accommodation have central heating?

- If you have central heating available, ✓ 'Yes' whether or not you use it.
- Central heating includes:
  - gas, oil or solid fuel central heating
  - night storage heaters
  - warm air heating
  - underfloor heating

☐ Yes, in some or all rooms

☐ No

H7 How many cars or vans are owned, or available for use, by one or more members of your household?

- Include any company car or van if available for private use.

☐ None

☐ One

☐ Two

☐ Three

☐ Four or more, *please write in number* ☐☐

H8 Does your household own or rent the accommodation?

- *✓ one box only.*

☐ Owns outright ► Go to **H10**

☐ Owns with a mortgage or loan ► Go to **H10**

☐ Pays part rent and part mortgage (shared ownership) ► Go to **H10**

☑ Rents ► Go to **H9**

☐ Lives here rent free ► Go to **H9**

H9 Who is your landlord?

☐ Council (Local Authority)

☐ Housing Association
Housing Co-operative
Charitable Trust
Registered Social Landlord

☐ Private landlord or letting agency

☐ Employer of a household member

☐ Relative or friend of a household member

☐ Other

H10 Please turn the page.

Household Members and their Relationships within the Household

- The example below shows how to provide the relationship information for John Smith, his wife (Mary) and their three children (Alison, Steven and James).
- In this example Steven's (Person 4) relationship to Person 1 is son, to Person 2 is son and to Person 3 is brother.

Name of Person 1
First name: JOHN
Surname: SMITH

ENTER NAME OF PERSON 1 ABOVE

Name of Person 2
First name: MARY
Surname: SMITH

Relationship of Person 2 to Person ➡	1
Husband or wife	☑
Partner	☐
Son or daughter	☐
Step-child	☐
Brother or sister	☐

Name of Person 3
First name: ALISON
Surname: SMITH

Relationship of Person 3 to Person ➡	1	2
Husband or wife	☐	☐
Partner	☐	☐
Son or daughter	☑	☑
Step-child	☐	☐
Brother or sister	☐	☐

- Use the same order and person numbers as in Table 1 (page 2), starting with Person 1.
- Print the name of each household member in the space at the top of each column.
- ✓ a box to show the relationship of each person to each of the other members of your household.
- Include relationship information for household members who require an Individual Form for privacy reasons. Questions on the following pages should be left blank for these people.

Name of Person 1
First name
Surname

ENTER NAME OF PERSON 1 ABOVE

Name of Person 2
First name
Surname

Relationship of Person 2 to Person ➡	1
Husband or wife	☐
Partner	☐
Son or daughter	☐
Step-child	☐
Brother or sister	☐
Mother or father	☐
Step-mother or step-father	☐
Grandchild	☐
Grandparent	☐
Other related	☐
Unrelated	☐

Name of Person 3
First name
Surname

Relationship of Person 3 to Person ➡	1	2
Husband or wife	☐	☐
Partner	☐	☐
Son or daughter	☐	☐
Step-child	☐	☐
Brother or sister	☐	☐
Mother or father	☐	☐
Step-mother or step-father	☐	☐
Grandchild	☐	☐
Grandparent	☐	☐
Other related	☐	☐
Unrelated	☐	☐

Name of Person 4

First name **STEVEN**
Surname **SMITH**

Relationship of Person 4 to Person →	1	2	3
Husband or wife	☐	☐	☐
Partner	☐	☐	☐
Son or daughter	☑	☑	☐
Step-child	☐	☐	☐
Brother or sister	☐	☐	☑

Name of Person 5

First name **JAMES**
Surname **SMITH**

Relationship of Person 5 to Person →	1	2	3	4
Husband or wife	☐	☐	☐	☐
Partner	☐	☐	☐	☐
Son or daughter	☑	☑	☐	☐
Step-child	☐	☐	☐	☐
Brother or sister	☐	☐	☑	☑

Name of Person 4

First name
Surname

Relationship of Person 4 to Person →	1	2	3
Husband or wife	☐	☐	☐
Partner	☐	☐	☐
Son or daughter	☐	☐	☐
Step-child	☐	☐	☐
Brother or sister	☐	☐	☐
Mother or father	☐	☐	☐
Step-mother or step-father	☐	☐	☐
Grandchild	☐	☐	☐
Grandparent	☐	☐	☐
Other related	☐	☐	☐
Unrelated	☐	☐	☐

Name of Person 5

First name
Surname

Relationship of Person 5 to Person →	1	2	3	4
Husband or wife	☐	☐	☐	☐
Partner	☐	☐	☐	☐
Son or daughter	☐	☐	☐	☐
Step-child	☐	☐	☐	☐
Brother or sister	☐	☐	☐	☐
Mother or father	☐	☐	☐	☐
Step-mother or step-father	☐	☐	☐	☐
Grandchild	☐	☐	☐	☐
Grandparent	☐	☐	☐	☐
Other related	☐	☐	☐	☐
Unrelated	☐	☐	☐	☐

Remaining questions should be answered by each member of your household in the same order as Table 1 (page 2 of this Form). Where a household member is completing an Individual Form for privacy reasons, the remaining questions for this person should be left blank.

Person 1

1 What is your name? (Person 1 in Table 1)

First name and surname

☐☐☐☐☐☐☐☐☐☐☐☐☐☐☐☐☐☐☐☐☐☐

2 What is your sex?

☐ Male ☐ Female

3 What is your date of birth?

Day ☐☐ Month ☐☐ Year ☐☐☐☐

4 What is your marital status (on 29 April 2001)?

☐ Single (never married)

☐ Married (first marriage)

☐ Re-married

☐ Separated (but still legally married)

☐ Divorced

☐ Widowed

5 Are you a schoolchild or student in full-time education?

☐ Yes ► Go to **6**

☐ No ► Go to **7**

6 Do you live at the address shown on the front of this form during the school, college or university term?

♦ Only answer this question if you have answered 'Yes' to Question 5.

☐ Yes, I live at this address during the school/college/university term

► Go to **7**

☐ No, I live elsewhere during the school/college/university term

► Go to **36**

7 What is your country of birth?

☐ England ☐ Wales

☐ Scotland

☐ Northern Ireland

☐ Republic of Ireland

☐ Elsewhere, *please write in the present name of the country*

☐☐☐☐☐☐☐☐☐☐

☐☐☐☐☐☐☐☐☐☐

8 What is your ethnic group?

♦ Choose ONE section from A to E, then ✓ the appropriate box to indicate your cultural background.

A White

☐ British ☐ Irish

☐ Any other White background, *please write in*

☐☐☐☐☐☐☐☐☐☐

☐☐☐☐☐☐☐☐☐☐

B Mixed

☐ White and Black Caribbean

☐ White and Black African

☐ White and Asian

☐ Any other Mixed background, *please write in*

☐☐☐☐☐☐☐☐☐☐

☐☐☐☐☐☐☐☐☐☐

C Asian or Asian British

☐ Indian ☐ Pakistani

☐ Bangladeshi

☐ Any other Asian background, *please write in*

☐☐☐☐☐☐☐☐☐☐

☐☐☐☐☐☐☐☐☐☐

D Black or Black British

☐ Caribbean ☐ African

☐ Any other Black background, *please write in*

☐☐☐☐☐☐☐☐☐☐

☐☐☐☐☐☐☐☐☐☐

E Chinese or other ethnic group

☐ Chinese

☐ Any other, *please write in*

☐☐☐☐☐☐☐☐☐☐

☐☐☐☐☐☐☐☐☐☐

9 This question is not applicable in England.

► Go to **10**

10 What is your religion?

♦ This question is voluntary.

♦ ✓ *one box only.*

☐ None

☐ Christian (including Church of England, Catholic, Protestant and all other Christian denominations)

☐ Buddhist

☐ [illegible]

☐ Jewish

☐ Muslim

☐ Sikh

☐ Any other religion, *please write in*

☐☐☐☐☐☐☐☐☐☐

☐☐☐☐☐☐☐☐☐☐

11 Over the last twelve months would you say your health has on the whole been:

☐ Good?

☐ Fairly good?

☐ Not good?

12 Do you look after, or give any help or support to family members, friends, neighbours or others because of:

- **long-term physical or mental ill-health or disability, or**
- **problems related to old age?**

♦ Do *not* count anything you do as part of your paid employment.

♦ ✓ time spent in a typical week.

☐ No

☐ Yes, 1 - 19 hours a week

☐ Yes, 20 - 49 hours a week

☐ Yes, 50+ hours a week

Person 1 - *continued*

13 Do you have any long-term illness, health problem or disability which limits your daily activities or the work you can do?

♦ Include problems which are due to old age.

☐ Yes ☐ No

14 What was your usual address one year ago?

♦ If you were a child at boarding school or a student one year ago, give the address at which you were living during the school/college/university term.

♦ For a child born after 29 April 2000, ✓ 'No usual address one year ago'.

☐ The address shown on the front of the form

☐ No usual address one year ago

☐ Elsewhere, *please write in below*

Postcode

15 If you are aged 16 to 74 ► Go to **16**

If you are aged 15 and under, or 75 and over ► Go to **36**

16 Which of these qualifications do you have?

♦ ✓ *all the qualifications that apply or, if not specified, the nearest equivalent*

☐ 1+ O levels/CSEs/GCSEs (any grades)

☐ 5+ O levels, 5+ CSEs (grade 1), 5+ GCSEs (grades A-C), School Certificate

☐ 1+ A levels/AS levels

☐ 2+ A levels, 4+ AS levels, Higher School Certificate

☐ First Degree (eg BA, BSc)

☐ Higher Degree (eg MA, PhD, PGCE, post-graduate certificates/diplomas)

☐ NVQ Level 1, Foundation GNVQ

☐ NVQ Level 2, Intermediate GNVQ

☐ NVQ Level 3, Advanced GNVQ

☐ NVQ Levels 4-5, HNC, HND

☐ Other Qualifications (eg City and Guilds, RSA/OCR, BTEC/Edexcel)

☐ No Qualifications

17 Do you have any of the following professional qualifications?

♦ ✓ *all the boxes that apply.*

☐ No Professional Qualifications

☐ Qualified Teacher Status (for schools)

☐ Qualified Medical Doctor

☐ Qualified Dentist

☐ Qualified Nurse, Midwife, Health Visitor

☐ Other Professional Qualifications

18 Last week, were you doing any work:

- **as an employee, or on a Government sponsored training scheme,**
- **as self-employed/freelance, or in your own/family business?**

♦ ✓ 'Yes' if away from work ill, on maternity leave, on holiday or temporarily laid off.

♦ ✓ 'Yes' for any paid work, including casual or temporary work, even if only for one hour.

♦ ✓ 'Yes' if you worked, paid or unpaid, in your own/family business.

☐ Yes ► Go to **24**

☐ No ► Go to **19**

19 Were you actively looking for any kind of paid work during the last 4 weeks?

☐ Yes ☐ No

20 If a job had been available last week, could you have started it within 2 weeks?

☐ Yes ☐ No

21 Last week, were you waiting to start a job already obtained?

☐ Yes ☐ No

22 Last week, were you any of the following?

♦ ✓ *all the boxes that apply.*

☐ Retired

☐ Student

☐ Looking after home/family

☐ Permanently sick/disabled

☐ None of the above

23 Have you ever worked?

☐ Yes, *please write in the year you last worked*

► Go to **24**

☐ No, have never worked

► Go to **36**

24 Answer the remaining questions for the *main* job you were doing last week, or if not working last week, your last *main* job.

♦ Your *main* job is the job in which you usually work the most hours.

25 Do (did) you work as an employee or are (were) you self-employed?

☐ Employee

☐ Self-employed with employees

☐ Self-employed/freelance without employees

26 How many people work (worked) for your employer at the place where you work (worked)?

♦ If you are (were) *self-employed,* ✓ to show how many people you employ (employed).

☐ 1 - 9 ☐ 10 - 24

☐ 25 - 499 ☐ 500 or more

Person 1 - *continued*

27 What is (was) the full title of your *main* job?

- For example, PRIMARY SCHOOL TEACHER, STATE REGISTERED NURSE, CAR MECHANIC, TELEVISION SERVICE ENGINEER, BENEFITS ASSISTANT.
- Civil Servants, Local Government Officers - give job title not grade or pay band.

28 Describe what you do (did) in your *main* job.

29 Do (did) you supervise any other employees?

- A supervisor or foreman is responsible for overseeing the work of other employees on a day-to-day basis.

☐ Yes ☐ No

30 What is (was) the business of your employer at the place where you work (worked)?

- For example, MAKING SHOES, REPAIRING CARS, SECONDARY EDUCATION, FOOD WHOLESALE, CLOTHING RETAIL, DOCTOR'S SURGERY.
- If you are (were) self-employed/freelance or have (had) your own business, what is (was) the nature of your business?
- Civil Servants, Local Government Officers - please specify your Department.

31 If you were working last week ► Go to 32

If you were not working last week ► Go to [illegible]

32 What is the full name of the organisation you work for in your *main* job?

- If you have your own business, write in the name.

☐ Self-employed/freelance ☐ Work for a private individual

33 What is the address of the place where you work in your *main* job?

- If you report to a depot, write in the depot address.

Postcode

☐ Mainly work at or from home ☐ Offshore installation

☐ No fixed place

34 How do you usually travel to work?

- ✓ *one box only.*
- ✓ the box for the longest part, *by distance*, of your usual journey to work.

☐ Work mainly at or from home

☐ Underground, metro, light rail, tram

☐ Train

☐ Bus, minibus or coach

☐ Motor cycle, scooter or moped

☐ Driving a car or van

☐ Passenger in a car or van

☐ Taxi

☐ Bicycle

☐ On foot

☐ Other

35 How many hours a week do you usually work in your *main* job?

- Answer to nearest whole hour.

Give average for last four weeks.

Number of hours worked a week

36 THERE ARE NO MORE QUESTIONS FOR PERSON 1.

- **Go to questions for Person 2.**
- **If there are no more people in your household you do not need to answer any more questions. Please leave the following pages blank.**
- **Remember to sign the Declaration on page 1.**

Subject Index